Stat Shot

The Ultimate Guide To Hockey Analytics

Rob Vollman
With Tom Awad
And Iain Fyffe

EasyRead Large

Copyright Page from the Original Book

Published by ECW Press
665 Gerrard Street East
Toronto, ON M4M 1Y2
416-694-3348 / info@ecwpress.com

Editor for the press: Michael Holmes
Cover design: Michel Vrana
Cover images: ice hockey icon set
© Askold Romanov/iStockPhoto
Interior images: Josh Smith

LIBRARY AND ARCHIVES CANADA CATALOGUING IN PUBLICATION

Vollman, Rob, author
Hockey abstract presents . . . stat shot : the ultimate guide to hockey analytics / Rob Vollman; with Tom Awad and Iain Fyffe.

ISSUED IN PRINT AND ELECTRONIC FORMATS.

1. Hockey—Statistics.
2. Hockey players—Statistics.

I. Title. II. Title: Stat shot.

GV847.V65 2016 796.96202'1 C2016-902353-2
C2016-902354-0

The publication of *Hockey Abstract Presents . . . Stat Shot* has been generously supported by the Government of Canada through the Canada Book Fund. *Ce livre est financé en partie par le gouvernement du Canada.* We also acknowledge the contribution of the Government of Ontario through the Ontario Book Publishing Tax Credit and the Ontario Media Development Corporation.

PRINTED AND BOUND IN CANADA

PRINTING: MARQUIS 5 4 3 2 1

TABLE OF CONTENTS

FOREWORD i
INTRODUCTION v
WHAT'S THE BEST WAY TO BUILD A TEAM? 1
WHAT DO A PLAYER'S JUNIOR NUMBERS TELL US? 97
WHO IS THE BEST FACEOFF SPECIALIST? 161
WHO IS THE BEST SHOT-BLOCKER? 202
WHO IS THE BEST HITTER? 216
WHO IS THE BEST PUCK STOPPER? 231
EVERYTHING YOU EVER WANTED TO KNOW ABOUT SHOT-BASED METRICS (BUT WERE AFRAID TO ASK) 317
WHAT WAS THE MOST ONE-SIDED TRADE OF ALL TIME? 394
QUESTIONS AND ANSWERS 442
CONCLUSION 459
GLOSSARY 460
ABOUT THE AUTHORS 468
BACK COVER MATERIAL 473

FOREWORD

by SEAN McINDOE
Author of The Best of Down Goes Brown
Former columnist at Grantland

I can still remember the excitement that came with the realization that hockey was on the verge of entering the analytics era. I'd watched as the Bill James–led sabermetrics movement swept through major league baseball fandom before eventually going mainstream around the turn of the century. As a diehard Maple Leafs fan going back over 30 years, I was eager to see the same scenario finally play out in my favourite sport. After all, the promise of analytics—more stats, better information, and ultimately a better understanding of the sport we love—is nearly irresistible to any true fan.

And sure enough, maybe five or six years ago, the first wave of hockey analytics started to show up on my radar. There were only two problems.

The first was that, back in those early days, good content could be hard to find. Oh, there was lots of it and the small community of fans who were doing most of the work knew where to find it, but to an average fan like me, it could be confusing to wade through random blogs, forums, and comments sections to figure out what was worth paying attention to.

The second and far more pressing problem: math is hard, and I'm not all that bright.

That first problem has largely disappeared, thanks to the work of many smart and dedicated fans, including the authors of the book you're now holding. You can now find engaging analytics work on virtually all of the major news and sports media sites, and the topic often comes up in broadcasts and highlight shows. More advanced work is being done on blogs and websites around the hockey world, supplemented by discussion and debate on social media. And, of course, NHL teams themselves have leapt enthusiastically onto the bandwagon, hiring up many of the brightest minds to inform their front office decision-making. In fact, just about the only place you couldn't find analytics was on your bookshelf, and now we've solved that issue too. These days, smart hockey talk is everywhere.

The second problem is a bit trickier. If you're a math lover who already gets the tougher concepts then you're all set. But if you're like me, and just calculating the tip on a $20 restaurant bill feels like a homework assignment, it can sometimes feel like you're slamming your head against a wall. I've found myself reading through 2,000 word posts about some particular newly developed stat, getting to the bottom where they lay out the results for certain key players, and then thinking "Wait, is a high number supposed to be good or bad?"

Some of that is by design—not every analytics article is meant for a general audience. A lot of the pieces are intended for fellow experts who'll engage in a form of peer review to make sure the work is solid. That's a good thing, and it's the sort of approach that will keep the movement headed in the right direction. But there's a certain skill needed to do good work that's also accessible to a wide audience.

That's where Rob Vollman comes in. I've been reading Rob's work for years, and there may be no better writer in the field when it comes to straddling that line between producing smart, important content while still making sure the average fan can get their head around it. There's nothing quite like getting to the end of a complex piece and realizing "Hey, I get this. It makes sense. I can actually use this."

That's the beauty of Rob's work in general, and of this book in particular. If you're just dipping a toe into the analytics waters for the first time, you'll be surprised at just how much of these "advanced" stats will end up making intuitive sense to you. If you're a novice, you'll encounter a mix of familiar concepts and brand-new ones that take the basic concepts further, or sometimes in surprising new directions. And if you're already an expert, well, you're probably not reading this foreword. You dove straight in, eager to see what sorts of new discoveries you'll find inside.

In any case, you'll come out the other side a smarter hockey fan. You'll learn which stats to follow and which to largely ignore. You'll be better able to predict what's most likely to happen in the future. And, maybe most important of all, you'll end up with a much more finely tuned BS-detector, one that will keep you from getting suckered in by common fallacies and sloppy thinking. You won't necessarily like all the lessons you learn—I'm still not quite over the ugly truth analytics taught me about first-line centre Tyler Bozak—but you'll be wiser for them.

The work of Rob and folks like him has undoubtedly made me a better hockey fan and, I hope, a much better hockey writer. Spend some time with this book, and they can do the same for you.

(You'll still be on your own for those restaurant bills though.)

INTRODUCTION

It was called the summer of analytics. The growing wave of interest in the field reached a tipping point early in 2014, with the successful, high-profile prediction of Toronto's late-season disaster followed by the triumph of Justin Williams and the Los Angeles Kings in the Stanley Cup Final, all of which prompted a flurry of interest in hockey analytics from NHL front offices and mainstream media.

After years of toiling away in anonymity, our hockey analytics community had a busy and exciting time that off-season. Of those analysts who had their work referenced in *Hockey Abstract 2014,* these eight were subsequently hired by NHL teams:

NHL FRONT OFFICE HIRING OF
HOCKEY ANALYTICS OUTSIDERS, 2014

DATE HIRED	TEAM	ANALYST
July 14	CAR	Eric Tulsky
August 5	EDM	Tyler Dellow
August 10	FLA	Brian Macdonald
August 19	TOR	Darryl Metcalf
August 19	TOR	Rob Pettapiece
September 29	undisclosed	Corey Sznajder
October 11	WSH	Timothy Barnes
October 20	SJS	Matt Pfeffer

In addition, Toronto hired Cam Charron on August 19th, an undisclosed club hired Dimitri Filipovic on September 29th, and New Jersey, perhaps kicking off the entire party, had hired Sunny Mehta as its director of hockey analytics on June 12th.

Of course, hockey statisticians had been hired in the past, but what made this special was the fact that teams were choosing outsiders for those roles. Not exclusively, of course, given the equally prominent hiring of analysts like Ian Anderson in Philadelphia and Kyle Dubas in Toronto, but bloggers and amateur pundits were certainly being pursued to a far greater extent than ever before, and teams were being far less secretive about it.

Picking up on the fact that this sport was getting ready to join baseball and football in publicly acknowledging and tentatively embracing these new views, mainstream media quickly followed suit, hiring many of the remaining analysts for various websites, newspapers, magazines, and radio shows. Most importantly, the NHL itself launched a multi-phase project in the 2014–15 season, introducing a flurry of statistics to its website that were innovated exclusively by outsiders and bloggers.

Even in my personal experience, the growing popularity of this field has been obvious. Not only did I get this very book deal after years of self-publishing, but I was regularly contacted with questions by about a dozen NHL organizations and several mainstream media organizations. My work was even featured in non-sports magazines like *Rolling Stone* and *Forbes.**

After over a decade of silence, interest in hockey analysts' work was growing rapidly.

[* Steve Lepore, "The NHL's Numbers Game: The Evolution of Hockey's Analytics Movement," *The Rolling Stone,* November 5, 2014, http://www.rollingstone.com/culture/features/the-nhl-numbers-gamethe-evolution-of-hockeys-analytics-movement-20141105; Jim Pagels, "Hockey Analytics Conferences Continue Growing, Following Path of Other Sports," *Forbes,* April 13, 2015, http://www.forbes.com/sites/jimpagels/2015/04/13/hockey-analytics-conferences-continue-growing-following-path-of-other-sports/.]

Since it's never too early to illustrate a point with a chart (and because it's only fair that I should be the first individual subjected to the same type of analysis that will soon be unleashed on all the NHL's players), the following chart shows the interest in my own brand of statistical analysis on radio and television, growing from about 10 appearances per year historically to about 50 in 2014 and 100 in 2015.

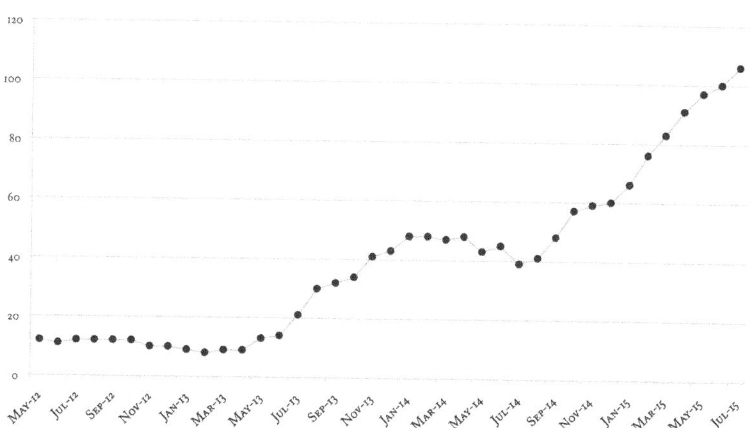

This same trend is typical of my fellow statistical analysts and across other forms of media, including frequent quotes and nontraditional statistics appearing in mainstream websites, newspapers, and magazines. Like it or not—and you'd presumably be holding this book only if you like it—statistical analysis has finally arrived to our sport to stay.

A Mixed Blessing

The surging popularity of hockey analytics comes with the good, the bad, and the ugly.

On one hand, it finally provides a platform for the growing number of brilliant analysts to share their work and to build on each other's developments, thereby rapidly advancing the field. On the flip side, it is challenging for mainstream fans to make sense of these new statistics, especially in how they can be applied to real hockey situations. For all the fantastic work that's out there, there are also studies that are confusing, incomplete, misleading, or just plain wrong. Add it up, and it can be a frustrating experience for those who are otherwise excited to dive into this world and enjoy the statistical analysis of our favourite sport as much as baseball fans do with theirs.

Fortunately, that's where *Hockey Abstract* comes in. Inspired by *Baseball Abstract,* which was written by Bill James from 1977 through 1988, this book is the third in a series that is intended to introduce modern statistical hockey analysis, along with its proper applications and limitations.

One of the great blessings of the surging popularity is that this book will be available, for the first time, on the shelves of leading book retailers, thanks to the tremendous support of our new publisher, ECW Press.

On the down side, the realities of the publishing business mean that we have to skip a season and that the data, examples, and references within are going to be a year old. Fortunately, this is not an annual yearbook whose usefulness expires with every new season. Much like James's famous works in baseball, *Hockey Abstract* is more about *how* to examine the sport and not about the actual numbers themselves. Our approach, along with most of our findings, is intended to be timeless.

For those readers who want up-to-date data, there will be an electronic download on the Hockey Abstract website.* This digital supplement will also include the James-inspired team essays, which leverage our considerable experience with *Hockey Prospectus* and *McKeen's* magazine. The website itself also includes plenty of additional data along with some online tools, almost all of which we are pleased to present to everyone at no charge.

[* Hockey Abstract, http://www.hockeyabstract.com.]

On behalf of all three authors, we deeply appreciate those who supported the previous two self-published books, which were available only on Amazon (and still are), and we are very excited about how this new deal can help us deliver an even better product and reach a brand-new audience, much as Bill James first did in 1982.

The Inspiration of Bill James

Despite the audacity of this book's title, I'm not hockey's Bill James. I'm not even sure there is one, but I'm hoping that someday there will be a young man or woman who is being inspired by this book.

James was my inspiration. Extremely well-worn copies of *Baseball Abstract* were among the only books I ever pulled off the shelf in my youth. Like James, I naturally viewed the game (and the world) through a statistical lens, and I loved breaking down and organizing every element of that sport in an objective fashion. I quickly applied many of the ideas he wrote about to hockey, including quality starts and league translations. I strove to emulate his style in my early years until developing the confidence to blend in more of my own.

That's why this is not a book that's about what we know about hockey but rather what we don't know. Most topics begin with a question, which is then explored without any preconception of what the answer should be. Virtually everything I have ever done in this field began with the words "I don't know." Consequently, this isn't a book about hockey stats; rather, it's a book about hockey that uses stats.

There are those who say that hockey analytics just aren't there yet, and that is a fair criticism. We certainly haven't caught up to baseball, a sport that, over its 30-year head start, has established what works and what doesn't (and to what extent), built a track record of proven results, and figured out how best to express these concepts and/or tie them to traditional analysis. It will take time to do that in our favourite sport, and books like these are an important part of that journey.

Philosophically, our view is that statistical hockey analysis is at its best when serving as a sober second thought to confirm or challenge what has been observed through traditional analysis and when being used to find players, teams, or situations that might have been missed and deserve a closer look. As such, there will always be new questions to answer, including some where the numbers can play only a secondary role.

There will also always be new hockey statistics. Just as in baseball, there are different kinds of stats with many different purposes. In general, they can be classified based on how well they describe what has occurred and how well they predict what will happen in the future. Some metrics are only meant for the former, while others include the latter component to some extent or another; we lean toward the latter. Like James, we try to validate our findings using as many different statistics and perspectives as possible, not just our own and not just the most mature ones or those currently held in the greatest esteem.

Given our mainstream audience, we also considered a third property of hockey statistics—their complexity. Metrics that are straightforward, are accessible, and make intuitive sense are going to be far more useful in the long run than those that are slightly more accurate but require a master's in statistics to understand or require someone else's explanation and interpretation. While this medium does allow us the space to provide such explanations and interpretations, it makes more sense to include the more straightforward versions as well and ultimately let the readers reach their own conclusions.

If there's one overall guiding philosophy, then it's to share our passion by having fun. Too frequently, statistical analysis is used to criticize and tear down teams, players, or other people's work—and James himself was certainly no exception. While it's perfectly reasonable to challenge a particular view or specific players, it's not really what makes this world the most fun nor the most useful. For us, the passion comes from exploring creative new ideas and perspectives and finding ways to better understand various players, teams, and situations. Ultimately, that's what magnifies our appreciation of this great sport.

- HOCKEY CALENDAR

A Third Book

It is highly rewarding to have you as an audience for our work. Both new and old readers may already be wondering what to expect from this third book. Basically, the first book was a primer that laid some foundations, the 2014 offering built on those principles to break new ground, and this third one brings it all together.

Most notably, each of the authors chose one important and ambitious topic, or "big chapter," that blends several concepts together. While previous books are certainly a useful companion to this new research, absolutely no previous knowledge is necessary—everything is explained.

Specifically, Tom wrote the definitive guide to individual possession statistics, Iain laid out how the careers of young prospects can be projected using their junior statistics, and I built a model of how to use statistical analysis to build a team in the salary cap era.

The next four chapters revolve around the classic bar-stool "who is the best" arguments that we so love. In the past, we've used statistics to find the best goalie, best defensive player, best playmaker, best goal scorer, best penalty-killer, best power-play specialist, and even the best coach and most undervalued player. This year we're hunting for the best faceoff specialist, best shot-blocker, best hitter, and best puck stopper.

We also like to include a sprinkle of history in every edition; this year we're taking an updated look at the 20 most lopsided trades in history.

A new book is a great opportunity to build on the past and to create a better product. As always, we gratefully review our reader feedback and strive to bring forward the best elements from the previous editions, integrate new ideas, and ultimately provide a familiar but superior overall experience.

With that in mind, those who enjoyed the previous two books can look forward to seeing the same basic philosophy and structure and all their favourite elements that made past editions fun to read, including the following:

- There is absolutely nothing within but all-new and original material.

- This is a meaty book that can be jumped into anywhere. *Hockey Abstract* includes a detailed table of contents and glossary.
- There are plenty of illustrations, examples, anecdotes, charts, and graphs to break things up, to help explain how a result was arrived at, and to translate those results into meaningful and timeless points.
- This book features the spirited collaboration of three authors who have worked together for over a decade and who seamlessly complement each other's work. According to readers, not only does this provide a nice mix,* but it avoids the perception that this perspective comes from a single gatekeeper.**

[* John Fischer, "Book Review: *Rob Vollman's Hockey Abstract 2014," In Lou We Trust* (blog), September 7, 2014, http://www.inlouwetrust.com/2014/9/7/6112495/book-review-rob-vollman-hockey-abstract-2014.]

[** J.J. from Kansas, "Book Review: *Rob Vollman's Hockey Abstract 2014," Winging It in Motown* (blog), August 19, 2014, http://www.wingingitinmotown.com/2014/8/19/6043393/book-review-rob-vollmanshockey-abstract-2014.]

- There is a wealth of interesting little discoveries and insights. For example, in the 2014 edition, readers enjoyed learning how scoring is up 15% in the second period because of the long change, that penalties are down 17% in the third period because refs are reluctant to call penalties, and that backup goalies have their stats negatively affected by getting the lion's share of back-to-back road games.

- New statistics. The 2014 edition introduced home-plate save percentage and "dirty rat" penalty minutes, and this edition presents new faceoff, goaltending, and shot-blocking statistics.
- There are still some careful challenges of controversial topics, which were a key attraction of the 2014 edition, including who should be in the Hall of Fame, a plunge into the shot-quality debate, questioning if enforcers are necessary, the blown Ottawa Senators prediction from the inaugural edition, and the goalie one-stat argument.
- The well-received Q&A chapter, which closes the book, will cover any remaining ground and provide any required updates to previous works.
- Finally, the warmest praise these books have received was all about being so much fun to read. This is not a hostile attack on teams, colleagues, or the mainstream, nor is it some kind of contest about proving who is clever. Our work is at its best when our deep love for what we do shines through.

Naturally, not all the responses to the previous editions were positive, and we have tried to acknowledge what can be improved with equal gratitude. For example, we tried to fill in more gaps this year, specifically related to how factors such as quality of competition, quality of teammates, and zone starts can affect possession numbers. Each of these concepts takes more of a centre stage in Tom's chapter on possession statistics.

Perhaps the toughest criticism I've seen was from a reader who felt that our last book was "hard work, but may as well be free."* In fairness, I partly understand his viewpoint and how things have changed since Bill James and *Baseball Abstract.* With all the great free content available online, and with free access to great statistical databases like *Behind the Net,* why buy a book?

[* John Fischer, "Book Review: *Rob Vollman's Hockey Abstract 2014," In Lou We Trust* (blog), September 7, 2014, http://www.inlouwetrust.com/2014/9/7/6112495/book-review-rob-vollman-hockey-abstract-2014.]

First of all, don't diminish the value of staying on top of the cutting-edge research and hard-to-find data all in one place. Some fans don't get the opportunity to hear more than the perspectives of a handful of the most prominent and/or local statistical analysts, which is why being kept up to date on the work of well over 50 different experts (and future front office managers) is, on its own, well worth the cover price.

Even for those who enjoy these free online articles as much as we do, they don't frequently allow the writer to go into any particular detail or to really explore where the underlying data and concepts came from, along with their proper application and limitations. Years ago, I remember exchanging messages with one particularly bright young man with an almost insatiable curiosity about stats like goals versus threshold (GVT). Several of us responded to his questions as best we could through emails, tweets, and links to articles, but until a book like this was complete, there was no single source that could quench his thirst. By then he had fallen back on the numbers he already knew and understood and had gone silent.

This may not exactly sound like a Shakespearean tragedy, but I was once a young man like that, and I don't like to think about all the joy I'd have missed out on if I hadn't found Bill James's books to help fuel my passion. For us authors, there is no greater joy than when we connect with someone who was in some way motivated or inspired by our work, and we can't wait to see what they contribute to our beloved sport. That's exactly why we do this.

Finally, we want more books like these, not just our own. Call us old-fashioned, but we like to walk into a bookstore and physically flip through the pages of statistical books the same way baseball fans can. There's not a lot of fame or money in this, but the support *Hockey Abstract* has enjoyed will soon make it possible for other writers and publishers to get involved, which brings me to the final sentiment we'd like to share before getting started.

A Special Thank You

The purpose of this introduction was to give everyone an idea of why we wrote this book and what to expect, along with an overview of our general approach to and philosophy analytics. But the most important words on these opening pages are to thank you for reading.

Writing these books has been a tremendously challenging, joyful, and rewarding experience thanks in large part to the tremendous support of our readers. That response really makes a big difference, both in the quality of our work and to all of us personally. While we do get a surge of enthusiasm every time we work with a front office or get attention from the mainstream media, our greatest passion is with our fellow fans.

That's why we have used some of the proceeds to help organize, promote, and attend grassroots hockey analytics conferences. These fan-held events have already been hosted in Edmonton, Calgary, Pittsburgh, Ottawa, Washington, Rochester, and Boston, with more to hopefully follow in the future. We really hope to meet you at an upcoming event, thank you for reading our ideas, and hear your ideas.

From one group of fans to another, thank you and please enjoy.

WHAT'S THE BEST WAY TO BUILD A TEAM?

by ROB VOLLMAN

Without question, the most ambitious topic to tackle with statistical hockey analysis is how to build a team. Not only is team management an extremely challenging subject, but many of its key concepts aren't exactly easy to explain in a meaningful and entertaining way. On the other hand, what's the point of a book like this if it shies away from this type of question?

The analysis here is entirely focused on the post-2005 salary cap era, when the dynamics of how teams are built completely changed. For example, the Chicago Blackhawks brilliantly assembled a dominant collection of talent on their way to the 2010 Stanley Cup, but the salary cap forced them to part ways with superstars like Antti Niemi, Andrew Ladd, Dustin Byfuglien, and Brian Campbell, not to mention useful secondary players like Kris Versteeg, Troy Brouwer, and Tomas Kopecky. Somehow Chicago was able to successfully manage its roster and remain competitive by replacing those players with rookies and other bargains, winning yet again in both 2013 and 2015, only to find themselves in exactly the same position they had been in five years earlier. At season's end,

Chicago had eight forwards, four defencemen, and two goaltenders under contract for a combined $65 million, leaving the team with just over $6 million to fill six to nine remaining roster spots. Once again, some excellent players had to go and were replaced by rookies and bargain-priced depth players.

The Blackhawks are an ideal case study for a guide to building a team in the salary cap era, which could actually be a topic for an entire book. The NHL is a dynamic market of very different players at various points in their careers, with ever-changing market inefficiencies and a collective bargaining agreement (CBA) chock full of both rules and exceptions. There are entry-level contracts, several different types of restricted and unrestricted free agents, and many different types of bonuses as well as trade deadlines, waivers, front-loaded contracts, buyouts, and special rules for players both young and old.

How can we sort all this out? The primary concept is to create a team-building model upon which all of these rules can be added. Its primary goal is to maximize the expected value of a team while staying within the team's total cap space and abiding by all the numerous rules and regulations of the most recent CBA.

For this model, the central requirement is a method of projecting a player's expected value over the life of his contract relative to his expected cost in cap space. Among other factors, this method will have to weigh each player's offensive and defensive contributions, to allow a comparison between players of different types and positions, to consider the scarcity of each type of player, to project each player's future by including some kind of age curve to make the distinction between up-and-coming players and declining veterans, and to account for a wide variety of additional factors, such as a player's acquisition cost and situations where the opinions of the scouts significantly disagree with the numbers. That may sound like an overwhelming project, and more than a little dry, but it actually makes perfect sense when everything is broken down into bite-sized pieces. It also sheds some fascinating light on our favourite teams and players on a case-by-case basis.

Before we begin, it's critical to note that what's being presented in the following pages isn't the *only* team-building model, nor is it the perfect one, but it will fully represent what every model needs to look like. If newer and better methods come along for any of its components in the future, such as a better way to measure player talent or a superior projection system, then these methods can be easily substituted for what is included here. Above all, think of everything that is presented in this chapter as more of a way of thinking about the problem, as opposed to being a definitive solution in and of itself.

As a bonus, the completed model will produce a list of general rules and guidelines that apply in the here and now, like how much cap space should be invested in goaltending, and some tricks to getting the most out of free agency. Some of these rules and guidelines will be timeless, whereas others are a result of market inefficiencies that exist only at the present moment but may no longer be valid in the future. That's why the process by which these guidelines are discovered will be of far more interest than the conclusions themselves. After all, Chicago didn't become dominant by following trends that others uncovered years ago; they dominated by discovering and exploiting new opportunities.

These discoveries will be referred to as guidelines, instead of strict rules, because we aren't dealing with fantasy hockey teams that are being built from scratch and in a static universe. A front office already has a set of players and contracts from which to start, along with instructions from ownership and requests from the coaching staff. Furthermore, the desired players won't always be available in trade, through free agency, or in the team's farm system, leaving teams to make the best decisions possible with their available resources. That's why no team could ever achieve the perfect model, even if such a thing exists. The most successful team-building process is a dynamic one, with organizations following a set of gradually changing guidelines to forever improve their team one step at a time.

Before unveiling the model and exploring the resulting rules and guidelines, let's first take a close look at the salary cap and some of the specific details that are most relevant to what we intend to build.

The Salary Cap

Creating a team-building model would be easier if the salary cap were just a single, predictable sum of money under which the total combined salaries of all the players had to remain—but that would just be *too* easy.

In practice, the NHL's salary cap has many rules, and each rule has many exceptions, most of which change with each new CBA. A player's age, the number of games in which he has played, and the number of seasons during which he has played at least a certain threshold of games are all factors that need to be carefully monitored and considered. To be honest, I'm convinced some of the new conditions were added just so the greater complexity would justify more jobs and higher salaries for the league's lawyers and agents. That's why it's essential for each team to have an expert in cap-related matters on staff, who manages the model at all times. There are a lot of intricate rules and regulations that can be land mines for the unaware, as well as potential market inefficiencies that can be exploited by those who know the finer details.

While I'm sure that a comprehensive account of the salary cap and all of its rules would be a real page-turner, it is thankfully far outside the scope of this book. There are, however, a few significant details with which we need to be familiar before we can build the model.

Starting at the beginning, the NHL salary cap was introduced during the 2005 lockout. Well, reintroduced, actually—there was an NHL salary cap in the pre–Original Six days. Today's cap is known as a *hard cap* because there is no allowance for going over. A team that has run out of cap space doesn't pay a penalty; it would simply have to play its remaining games with fewer players, as the Calgary Flames did late in the 2008–09 season.[1]

---

1 There is a little-known new rule in the 2013 NHLPA CBA that allows a team that has dressed a short bench for two consecutive games some emergency replacements, who are paid no more than the league minimum plus $100,000,

The NHL's hard cap varies from year to year and quite unpredictably. It is calculated as a percentage of the NHL's revenues from the previous season, with a current minimum of $64.3 million. Initially, 54% of all hockey-related revenues went to the players, which increased to 57% in the 2013 agreement.

NHL SALARY CAP, 2005-06 TO 2015-16

SEASON	SALARY CAP	CHANGE
2005–06	$39,000,000	New
2006–07	$44,000,000	+12.8%
2007–08	$50,300,000	+14.3%
2008–09	$56,700,000	+12.7%
2009–10	$56,800,000	+0.1%
2010–11	$59,400,000	+4.6%
2011–12	$64,300,000	+7.6%
2012–13	$70,200,000	+9.2%
2013–14	$64,300,000	-8.4%
2014–15	$69,000,000	+7.3%
2015–16	$71,400,000	+3.5%

The salary cap for the 2015–16 season is $71.4 million, which is almost twice what it was when it was first introduced, for the 2005–06 season. That quickly rising salary cap is like a life raft for poor general managers. Deals that really don't make sense today could make sense down the line, as the deal's total percentage of a team's overall cap space decreases over time. Once that cap ceiling starts to stabilize, the teams with the most effective models will truly

---

which is yet one more reason for teams to have a cap expert on staff!

have the greatest advantage over the teams who frequently overpay.

Individually, there is also a player limit of 20% of the team's overall cap, or $14.28 million for the 2015–16 season. At the time of writing, the highest individual cap hit is $10.5 million for Chicago's Patrick Kane and Jonathan Toews.

Remember that it's the cap hit that must remain below $14.28 million, not the salary. A player's cap hit is calculated as his average salary over the length of his current contract, so adding some lower-paying seasons at the end of the deal can reduce the annual cap hit. Kane and Toews are actually being paid $13.8 million in the 2015–16 season, for instance, but only $6.9 million in their final two seasons, which is the legal minimum of 50% of the highest-paid year. This is known as a front-loaded deal, and it's a perfectly common, legal, and legitimate way to reduce a player's annual cap hit—even if he retires prior to playing out those final, low-paying seasons.

But watch out—the retired player's cap hit will continue to count if the contract went into effect after he turned 35. This is just one of the many points written in fine print (to create more jobs). Even without this clause, a veteran who plays out one of the lower-priced seasons may be frustrated if his subsequent contract doesn't include some extra compensation for that. This was the case with Daniel Alfredsson, a career-long Senator who spent his final season with Detroit after being displeased when he wasn't sufficiently rewarded for playing out a $1 million season with Ottawa in 2012–13.

As for term, there's also a maximum contract length of seven seasons, with an additional year allowed if the player is re-signing with the same team. Once again, there are some special job-creating exceptions. Players who sign their first NHL contracts, which are called entry-level contracts (ELCs), are limited to three-year deals or less, depending on their age. Furthermore, these players have their salaries capped at $925,000 plus an additional signing bonus capped at 10% of their initial salary. Even with additional performance bonuses, which can reach $2 million, ELCs are obviously the most affordable types of contracts and should be timed carefully to maximize a player's value.

These, and all other types of bonuses, *do* count against the cap. However, teams are allowed to go over their salary cap on performance-related bonuses, which may or may not occur, and carry them over into the next year. The down side of this practice is that it could leave a team with less cap space with which to work the following season. This was exactly the case for the Boston Bruins, who had $4.2 million of their 2014–15 cap space used up by a bonus earned by Jarome Iginla the previous season, a player who had since moved on to the Colorado Avalanche. Oh, and the Bruins missed the playoffs by two points that year. Doh!

There is also a cap floor, incidentally, which can be safely ignored in any model designed to produce champions. It's the individual league minimum that's more relevant to the model, because that's the cost of a replacement-level player. Even if a player isn't worth it, every NHLer must be paid this league minimum, which will increase far less gradually after the 2016–17 season.

INDIVIDUAL LEAGUE MINIMUM SALARY, 2005–06 TO 2021–22

SEASON	LEAGUE MIN	SEASON	LEAGUE MIN
2005–06	$450,000	2014–15	$550,000
2006–07	$450,000	2015–16	$575,000
2007–08	$475,000	2016–17	$575,000
2008–09	$475,000	2017–18	$650,000
2009–10	$500,000	2018–19	$650,000
2010–11	$500,000	2019–20	$700,000
2011–12	$525,000	2020–21	$700,000
2012–13	$525,000	2021–22	$750,000
2013–14	$550,000		

What does this mean for the model? Since a team dresses 20 players for each game, 18 skaters and two goalies, even a team of replacement-level players will use up $11.5 million of the team's total cap hit (in 2015–16 and 2016–17). Teams are allowed three additional players on their active roster, pushing the total up to just over $13.2 million. That means that our model is all about how a team should invest its remaining $58.2 million in cap space, or roughly $2.53 million per player. Signing too many players for a greater sum of money could expose teams to some risky situations and send them in search of cap relief.

Cap Relief

What can teams do with a bad contract? Beyond the previously discussed front-loaded deals, here are a few ways teams can achieve a little bit of cap relief.

1. SEND A PLAYER TO THE MINORS

Historically, useful but high-priced players were sent to the minor leagues, more specifically the American Hockey League (AHL), in order to avoid the player's cap hit. New York's Wade Redden was one prominent example in 2010, along with Sheldon Souray, Jeff Finger, and Mike Commodore in the following seasons.

As of the 2013 CBA, this no longer works for any player who is on a one-way contract (one that pays him the same salary regardless of where he plays), which is the case with most NHL regulars. In this case, the cap relief is limited to a maximum of $375,000 over and above the league minimum salary.

That's not much cap relief at all, but it does explain why those making around $950,000 are more likely to be sent to the AHL by cap-strapped teams, since a higher-paid player doesn't offer any further cap relief. And, as before, there is an exception for players who were 35 or older at the time their contract took effect, who provide no cap relief at all, regardless of where (or if) they play.

For added risk, NHL players sent to the minors, either to reduce the team's cap hit or because they aren't contributing, must clear waivers. That process allows every other NHL team the option of assuming that player's contract in return for an almost trivial monetary compensation for those rights. Prior to the new 2013 CBA, players also had to clear waivers upon their re-entry to the NHL, at which point they could be claimed by other teams for half the annual cap hit.

While the discontinuation of that particular clause might have killed some of those lawyer jobs, there is still a rather complicated set of regulations that exempt players under age 25 from having to clear waivers at all, based on when they played their first season and their number of career games played.

Though all of the exact details aren't necessary to build our model, teams must keep a keen eye on every player's status, especially as they approach one of those cut-offs. For example, one key threshold for rookies is 11 NHL games, at which point a player is defined as having played his first NHL season for this and several other rules. That's why you'll find that many first-time NHLers are limited to only 10 games in their first season.

2. LONG-TERM INJURED RESERVE

If a player is expected to miss at least 10 games over at least 24 days, he is eligible to be placed on long-term injured reserve (LTIR). While that player's cap hit still counts, the combined cap hits of any replacement players don't count toward the team's cap, except for any portion that is over and above the injured player's own cap hit. These injury replacements are also exempt from the waiver-clearing process described earlier.

Remember that only long-term injuries can provide any kind of cap relief. Therefore, it is improper to consider a player's cap hit on a per-game basis, since there is no relief for shorter and more occasional injuries, and certainly none for healthy scratches.

Since the cap relief applies only to replacement players, some teams trying to reach the cap floor seek out such players on purpose, like Marc Savard for the Florida Panthers or Chris Pronger for the Arizona Coyotes. On the flip side, some teams don't care about the underlying salaries and just want the cap relief, which leads to bizarre trades like Toronto acquiring the injured Nathan Horton from the Columbus Blue Jackets, to free up the cap space, in exchange for the otherwise useful but overpaid David Clarkson. This may give us fans a chuckle, but I wonder how it must feel to be the guy actually traded for someone who isn't playing anymore ... It's probably a little bit like high school felt for me.

3. RETAINING CAP SPACE IN TRADE

It isn't always easy to find a trading partner willing to take on a bad contract, with those aforementioned exceptions, which is why there is the option to retain a portion of a player's cap hit when making a deal. As of the 2013 CBA, teams can retain up to half of a player's cap hit when trading him, to a maximum of three contracts per team and totalling no more than 15% of the team's cap space. Entering the 2015–16 season, for example, the Toronto Maple Leafs are retaining portions of Carl Gunnarsson's and Phil Kessel's salaries, leaving space for one more, while Carolina is similarly retaining salary for Winnipeg's Jay Harrison and New Jersey's Tuomo Ruutu.

Retaining salary can make it easier for a team to clear up most of a bad contract's cap space and can sometimes come with other advantages. The Arizona Coyotes, for example, received an excellent collection of prospects and picks at the 2015 trade deadline, when they agreed to retain half of Keith Yandle's salary upon trading him to the cap-strapped New York Rangers. To an extent, that makes cap space essentially an asset that can be exchanged like any other.

4. BUYING OUT THE CONTRACT

The most common way that a player's cap hit can be reduced is by buying out his contract, thereby making him an unrestricted free agent. In this case, the team will continue to carry a cap hit that is two-thirds of the player's total multi-year deal, but it will be spread out over twice as many seasons. If the player is 26 or younger, it is further reduced to only one-third of the original deal's total cap hit.

Using the Toronto Maple Leafs as the example once again, they agreed to carry a combined cap hit of $5.33 million over four seasons, for an average of $1.33 million per season, when they bought out the final two years of Tim Gleason's contract, which was scheduled to carry an annual cap hit of $4 million.

Gleason's buyout made sense only if the Leafs felt that they could get a better player with the remaining $2.67 million of cap space, in order to justify dealing with $1.33 million of so-called dead money for four seasons. Given that Gleason ultimately signed for $1.2 million in Carolina, it stands to reason that the team was indeed better off with the extra cap space than with Gleason's services. And yes, I have noticed that I've used Toronto as an example of bad contracts as frequently as I have used Chicago for good ones—but it's unintentional, I assure you.

5. RETIREMENTS

Finally, retirements end a player's cap hit, but with one notable exception. If the player turned 35 prior to when his contract began, then his cap hit will continue to apply for the entire original duration of the deal. This so-called Mogilny rule means that the cap hit of the older veterans will count whether that player is active or not. If such a player retires or leaves for the KHL, buying him out is the only way to free up (a portion of) that cap space. Such contracts should therefore be considered very carefully.

Summing it all up, teams can achieve a certain degree of cap relief using some combination of front-loaded contracts and these five methods, but it can result in a lot of so-called dead money down the road. High cap hits at the end of a front-loaded contract, buyouts, retained salaries, and the previously mentioned carried-over bonuses could result in a team having a lot less cap space with which to work. Dead money should be carefully monitored, and it's a key variable in a team's salary cap model.

The Hockey Abstract Team-Building Model

The first rule of hockey analytics is that winning is what matters, and the second rule is that goals scored and allowed are the only factors that affect winning. If that sounds familiar, then you have probably heard of Alan Ryder, who crafted 10 such rules back in 2008.*

[* Alan Ryder, "The Ten Laws of Hockey Analytics," *Hockey Analytics* (blog), January 2008, http://hockeyanalytics.com/2008/01/the-ten-laws-of-hockey-analytics/.]

This means that there is no single best way to build a team. Whether a team is built around scoring goals or preventing them, whether it's based on a few franchise players or a well-balanced lineup, or whether it is linked to elite goaltending or generational scorers, goals are goals. There are, however, easier and safer ways to build that team.

Managing a team's roster in the salary cap era is somewhat like managing a financial portfolio. Although it's a simplistic and insensitive way of visualizing the task, the players can be thought of as stocks, and the goal is to manage the performance of these limited resources. That's why the model requires a way to measure a player's value in a very specific way. It must be expressed relative to a replacement-level player, in terms of cap space, and it must be projected over the entire duration of a player's deal. This requires using the player's history, comparable players, and some kind of age curve to calculate each individual's expected value.

Projecting Expected Player Value

The central component of a team-building model is a statistic that will measure a player's value, in terms of cap space, and project it up to eight seasons in the future. On the whole, that is no easy task, but it becomes far less daunting when that entire process is broken down into single, manageable steps.

There are a lot of ways of going about this, some of which may prove simpler or more useful than mine. That's why I'll be generously referencing and summarizing the work of others, to provide those additional perspectives and detail.

I'll also walk through how you can complete each of these tasks for yourself, in case you have your own ideas about how to fine-tune the approach or wish to update the model in the future, as circumstances change. In essence, this will be like a cooking recipe, except that the step-by-step instructions will include the purpose of each task and what happens if you increase or decrease each ingredient or select a substitute. Let's begin.

1. CAPTURING PLAYER VALUE IN A SINGLE METRIC

The first ingredient is a way to measure the performance of every player with a single number.

Years ago in the NHL, and still today in other hockey leagues, the only way to accomplish this objective was with some combination of points and plus/minus, but we have come a long way since then. Hockey may not have progressed quite as far as baseball in being able to measure all of a player's different types of contributions in a single metric, but there are several modern options available from which to choose, including, in chronological order:

- Points allocations (PA), the first of its kind, introduced in 2002 by my co-author Iain Fyffe.*

[* Iain Fyffe, "Point Allocation," *Hockey Historysis* (blog), April 9, 2002, http://hockeyhistorysis.blogspot.ca/2014/09/puckerings-archive-point-allocation-09.html.]

- Goals versus threshold (GVT), introduced in 2003 by my other co-author, Tom Awad. (Apparently this was my unwritten prerequisite for being a contributor to this book.)*

[* Tom Awad, "Numbers on Ice: Understanding GVT, Part 1," *Hockey Prospectus* (blog), July 30, 2009, http://www.hockeyprospectus.com/puck/article.php?articleid=233.]

- Player contributions (PC), also introduced in 2003, by Alan Ryder.*

[* Alan Ryder, "Player Contribution," *Hockey Analytics* (blog), August 2003, www.hockeyanalytics.com/Research.../Player_Contribution_System.pdf.]

- DeltaSOT, the first shot-based and context-adjusted option, which Tom introduced in 2010.*

[* Tom Awad, "Numbers on Ice, Plus/Minus and Corsi Have a Baby," *Hockey Prospectus* (blog), January 21, 2010, http://www.hockeyprospectus.com/puck/article.php?articleid=436.]

- Point shares (PS), introduced by *Hockey Reference's* founder, Justin Kubatko, in 2011 and inspired by DeltaSOT, PA, GVT, and PC.*

[* Neil Paine, "Point Shares," *Hockey Reference* (blog), March 11, 2011, http://www.hockey-reference.com/blog/?p=101.]

- Total hockey rating (THoR), the second shot-based metric, introduced by Mike Schuckers and James Curro in 2013.*

[* Michael Shuckers and Jim Curro, "Total Hockey Ratings (THoR)," Statistical Sports Consulting, March 1, 2013, accessed April 9, 2015, http://statsportsconsulting.com/main/wp-content/uploads/Schuckers_Curro_MIT_Sloan_THoR.pdf.]

- dCorsi, the most recent of the shot-based metrics, introduced by Steve Burtch in 2014.*

[* Steve Burtch, "dCorsi – Introductions," *NHL Numbers* (blog), July 19, 2014, http://nhlnumbers.com/2014/7/19/dcorsi-introductions.]

- Wins above replacement (WAR), which is a GVT-like statistic calculated completely differently, introduced by Andrew Thomas and Sam Ventura in 2015.*

[* A.C. Thomas, "The Road to WAR Series," *War on Ice* (blog), April 8, 2015, http://blog.war-on-ice.com/the-road-to-war-series-index/.]

If none of these are to a team's liking, there is always the option of developing a new catch-all statistic, a process that was explained in the inaugural edition of *Hockey Abstract,* along with the details of all but the most recent metrics listed here.[2] In fact, I know of a few front offices that have already gone this route, in some cases many years ago.

In this book we'll use GVT because of its longer and more established history, its use elsewhere in this and previous books, and the accessibility of the data, but, mostly, because I'm pretty sure I'd get an earful from Tom if I went with one of the others.

While a goal-oriented, results-based statistic like GVT will serve nicely for forwards and goalies, I'm not as comfortable using it to measure the effectiveness of defencemen. Over the years we have learned from several analysts, most recently from Domenic Galamini, that defencemen have very little influence over their team's on-ice shooting and save percentages, and

---

2 Rob Vollman, "Catch-all Statistics," in Hockey Abstract (author, 2013), 143–159.

therefore over their team's on-ice scoring.* Quite frankly, it's a little unfair to measure their value around goal-based factors that are beyond their control. That's why possession-based statistics are far more appropriate for defencemen than the goal-based GVT, especially those that are adjusted for contextual factors like a defender's linemates, opponents, manpower situation, the zones in which he's used, and so on. To avoid another earful, I've chosen one of Tom's other statistics as a substitute, deltaSOT, which he has kindly modified for our purposes in a fashion compliant with the additional points that follow.

[* Domenic Galamini, "Possession Isn't Everything ... With the Exception of Defencemen," *Own the Puck* (blog), December 8, 2014, http://ownthepuck.com/2014/12/08/possession-isnt-everything-with-theexception-of-defensemen/.]

The beauty of this team-building model is that it can be constructed to use any of the other options instead (assuming Tom doesn't have your phone number) or even a superior catch-all statistic that is developed in the future. Each option has its own strengths and weaknesses, but the only key requirements are that the chosen metric measures all of a player's contributions in a single number and that it can be calculated relative to a player's cap hit.

2. ESTABLISHING REPLACEMENT LEVEL

When a player leaves an organization, for whatever reason, the team doesn't play a man short—the departing player is immediately replaced with someone else. Similarly, a newly acquired player will push his weakest new teammate out of the lineup. That's why a player's value shouldn't be measured relative to zero or the league average but to the difference between him and the next-best available hockey player at his position. That's a concept generally referred to as replacement level.

Calculating a player's value relative to a replacement-level player is the most common way for a catch-all metric to take scarcity into account. That means even a theoretical player of constant abilities would have an overall value that fluctuates from year to year with the availability of alternatives. For example, there are a lot of great defencemen available some seasons, while there have also been years where many of them were hurt, retired, left for the KHL, or struggled through vicious slumps. A blueliner's value would definitely go up in those latter cases.

So how is replacement level determined? In its simplest form, it is a bar that is established at whatever level the best applicable non-NHL player is performing. For example, it could be considered the 61st-best goalie, the 181st-best defenceman, the

361st-best forward excluding prospects, or, depending on the context, the 31st-best starting goalie or the 121st-best top-four defenceman. These can be determined using team depth charts, cap hits, assigned ice time, or one of the aforementioned statistics in a raw form.

In practice, it's not nearly that simple to calculate the league's true replacement level. If a starting goalie is lost, most of his starts go to the backup, not the AHLer who gets called up. Likewise, a top-pairing defenceman plays about 25 minutes a game, many of which get divided among his teammates in his absence, leaving only about 15 minutes to trickle down to his replacement. That drops even lower for forwards, as a newly promoted fourth-line option would be lucky to get to play more than nine minutes a game. In short, comparing a player directly to a replacement-level option is actually overstating the problem in most situations.

Not only is replacement level contextual, it is also dynamic. As a population of a certain type of player gets richer, that replacement level goes up, since the number of opportunities remains fixed. Likewise, replacement level goes down when significant numbers in that group get hurt, retire, or otherwise leave the available pool because teams still need the same number of players every year.

And yes, replacement level would most certainly drop during the next NHL expansion, when the number of openings increases. How do you think Teemu Selanne and Alexander Mogilny each scored 76 goals in 1992–93? This variable nature of replacement level is why it's important to recalculate every season, in order to keep the overall team-building model up to date.

In the case of this particular model, Tom has defined replacement level for GVT and deltaSOT as being about 75% of the league average, taking all factors into account. That is by no means a universal consensus, as Alan Ryder's player contributions, for example, adopts the concept of a marginal player based on 58% of the league average.* GVT presents a player's value not in absolute terms and not relative to the league average but in terms of how well the team would perform in his absence. That is a key concept and exactly what the VT stands for in GVT (i.e., versus a threshold).

[* Alan Ryder, "Player Contribution," *Hockey Analytics,* August 2003, accessed April 9, 2015, www.hockeyanalytics.com/Research_files/Player_Contribution_System.pdf.]

For example, Sidney Crosby's 20.4 GVT in 2014–15 measures his value relative to how the Penguins may have done without him. Without Crosby, players like Evgeni Malkin, David Perron, and Chris Kunitz would have been leaned on more heavily, while Blake Comeau and Nick Spaling would have moved up the depth chart, and someone like Jayson Megna or Andrew Ebbett would have secured a regular fourth-line NHL job alongside Craig Adams. However coach Mike Johnston chose to manage his absence, the end result would be that Pittsburgh's goal differential would decrease by 20.4 goals that year—presumably mostly in lost scoring, in this specific case.

Does 20.4 goals seem realistic, even as an upper bound? As a sober second thought (not that I've been drinking), consider the results Gabriel Desjardins discovered when he looked at all the NHL teams between 2002–03 and 2007–08 that went without one of their top players for a significant length of time. A season-long absence of this mixed group of top-line forwards yielded a net effect of roughly nine goals.* That's not atypical of a top-line forward's GVT and not out of line with the 20-goal value of a franchise player like Crosby.

[* Gabriel Desjardins, "Replacement Level: How Many Wins Do Injuries Cost?," *Behind the Net* (blog), April 17, 2008, http://behindthenet.ca/blog/2008/04/replacement-level-how-many-wins-do.html.]

While there is yet to be a consensus on the precise level of a replacement player, including the methods by which it should be determined and/or calculated, we can estimate that it is somewhere between 58% and 75% of the league average in today's NHL. Although individual team circumstances may vary, we're using the upper bound of that range in our model—and one that is recalculated each season.

3. PROJECTING THE FUTURE

Thus far, we have studied metrics that measure how effective a player was in the past, not how good he is today nor how well he's expected to perform in the future. Therefore, this model needs a way of converting that historical information into an estimate of a player's expected performance this season and over the remaining years in his entire contract.

The basic blueprint for predicting a player's upcoming season from his past is the Marcel method, introduced by prominent baseball and hockey analyst Tom Tango back in 2005. It's a three-step process that

1. starts with a weighted average of a player's three most recent seasons;
2. regresses that player's performance toward the league mean, based on games played (we'll explain why and how in a moment); and
3. applies an age adjustment to account for developing rookies and declining veterans.

With regard to the first step, Tango proposes a weighting of 5-4-3, while I prefer a 4-2-1 approach that assumes that every season's data has twice the predictive power as the season previous. In this case, that means multiplying a player's 2014–15 GVT by four, adding his 2013–14 GVT multiplied by two, and finally adding his 2012–13 GVT (scaled to an 82-game schedule) straight up before dividing the entire sum by seven.

While closer to it, this end result is not an accurate reflection of a player's actual present-day skill, because an observed performance is partly the result of the player's underlying true talent and partly a result of random variation. That is, when Crosby scored 84 points in 77 games in 2014–15, it was mostly because of his own incredible talent but also because of random chance. After all, he could have just as easily finished with as few as 70 points or as many as 100 or more.

Raw statistics can be perfectly suitable for explaining the past, but whenever the past is going to be used to predict the future, historical data should be adjusted to remove the effect of random variation, leaving only the reproducible skill component. To do otherwise is to waste your time.

So how much random variation is there in a particular player statistic? That all depends on how repeatable the player's performances have been. Traditionally, that's estimated by dividing the available data in two and seeing how well the two halves correlate.

- If the two halves have no correlation whatsoever, then it's a completely random event, and we should simply expect the league average in the future.
- If the two halves correlate very little, then we should assume that future production will be far

closer to the league average than to the observed results.

- If the two halves have a strong correlation, then we should assume that the future performance will be closer to the observed results than to the league average.
- If the two halves correlate perfectly, then we can assume that random variation plays no role at all, and we can use the observed results as a basis for a projection, with no regression whatsoever.

Bear in mind that this isn't a hockey concept, but one of basic statistics that applies equally to everyone and everything. You can go outside and measure yourself taking long jumps, for example, and figure out how much random variation is involved by comparing your odd-numbered jumps to your even-numbered jumps. You can even figure out at what point skill has more influence on your results than random chance. Or, I suppose, you can do something more productive with your life than we have and go take in a show or take your special someone out dancing. To each their own.

Getting back on the topic of this model, I calculated the correlation for GVT's three-year weighted average. Between the 2005–06 and 2008–09 seasons, there were 647 forwards who played in those four consecutive seasons. The correlation between a weighted average of their GVT through their first three seasons and that fourth season was a solid 0.65 (on a scale of 0 to 1), meaning that the observed performances involved mostly skill.

There is a still a noticeable element of random variation, which can be addressed by regressing everyone's data toward the league average by one minus the square root of the correlation, which gives us 19%.[3] Again, that is not a hockey concept but a long-standing formula in the world of statistics that was established by Sir Francis Galton well over a century ago and without any knowledge of what "hockey" even is.

These types of concepts can make a lot more sense with an example. Let's use Chicago's Jonathan Toews, who posted a GVT of 18.9 in 2014–15, 19.1 in 2013–14, and 15.7 in 2012–13 (pro-rated to an 82-game season). Multiplying 18.9 by four, multiplying 19.1 by two, and taking 15.7 straight up and dividing the total by seven results in an estimate of 18.5. To remove the potential impact of random variation, we add 81% of his 18.5 to 19% of a league-average GVT of 4.9 to yield a GVT of 15.9. That's the number we would use in the model as an estimate of his current-day performance.

One final technicality is to take the sample size of games played into account. Consider a simple statistic, like points per game and the lockout-shortened

---

3 The square root of 0.65 is 0.81, and 1 minus 0.81 is 0.19, or 19%.

2012–13 season. Toronto's Joffrey Lupul scored 18 points in 16 games, or 1.13 points per game. Obviously that result was a combination of both skill and luck, especially compared to Alex Ovechkin, who scored at roughly the same rate that year over a sample three times that size. This data was far more reliable in predicting the future for Ovechkin, who scored 1.01 points per game the following season, than for Lupul, who managed 0.64. Simply put, the data should be regressed toward the league average to a greater extent based on how limited the sample size is.

Finally, bear in mind that I've presented only the simplest possible form of statistical regression. Different and/or more sophisticated approaches can be adopted instead, especially ones that don't make the same assumption that NHL players fit a normal distribution curve. The key concept is that historical data should not be taken at strict face value when used to project the future. Random variation must first be quantified and removed using statistical regression, and it's really not as daunting a task as you may have feared.

4. ADJUSTING FOR AGE

Player performance isn't static. Younger players will develop, older players will decline, and the model can take that into account by incorporating an age curve.

For example, on July 9, 2014, Chicago re-signed its two superstars, Jonathan Toews and Patrick Kane, to identical eight-year extensions that carry annual cap hits of $10.5 million. While these are doubtlessly worthwhile contracts in 2015–16, for players at age 27, will they remain so in the 2022–23 season, with players at age 34? If not, will that late decline be bad enough that it cancels out today's early benefits? While that's essentially unpredictable on an individual basis, age curves can be used to figure out when and by how much players will decline, on average.

Building age curves is a little bit trickier than it may appear to be at first glance. Indeed, my first pass through this section was long enough to be a full chapter in its own right. Even the venerable hockey statistician Alan Ryder, who I was fortunate enough to have review my work, had to take a break halfway through to get some air and walk his dog.

Given the dry complexity of the topic, I had to find a way to make this section a little bit snappier and decided to simply leverage what my baseball colleagues had discovered years ago. After all, statistical baseball analysts have about a 20-year head start on us, during which time they have explored how athletes age in considerable detail. Consequently, they have found solutions for a number of problems, such as survivorship bias, selection bias, and annual changes in league scoring levels, all of which can also be applied to age curves in hockey.

Let's back it up and briefly start from scratch. At first glance, the obvious way to build an age curve is to simply calculate the league average for the target statistic, grouped by age. That's how Steve Burtch initially built an age curve for goalies, basing it on the average league save percentage for each age.* However, he quickly noticed that goaltending statistics didn't appear to drop with age. Why? Because any goalies who dropped below replacement level were demoted to the AHL or retired, so his calculation only ever included the great goalies who continued to play well enough to keep their jobs. Indeed, about a third of the goalies in his age-35 bucket were current or future Hall of Famers, far more than we'd expect from the age-25 group.

[* Steve Burtch, "Roberto Luongo Is the Best Option to Have in Goal," *Pension Plan Puppets* (blog), December 6, 2012, http://www.pensionplanpuppets.com/2012/12/6/3565650/roberto-luongo-is-the-bestoption-the-leafs-have-in-goal.]

Survivorship bias, as this is known, can sneak into an age curve in any number of ways. For example, Eric Tulsky ran into it when studying the effect of age on power-play scoring rates.* Since virtually any decent offensive player will get the opportunity to work with the man advantage at age 25 but only the truly gifted will continue to enjoy such assignments in their twilight seasons, power-play scoring rates will appear to increase for players well into the 30s, as the mediocre players drop off and only incomparable talents like Jaromir Jagr and Teemu Selanne remain, or "survive."

[* Eric Tulsky, "Time Spares No Man, Except Maybe the PP Specialist," *Broad Street Hockey* (blog), June 21, 2014, http://www.broadstreethockey.com/2013/6/21/4452220/power-play-points-aging-curve-knuble.]

One way around survivorship bias is to find a way to actually include the non-survivors in the calculation, much as Gabriel Desjardins once did by including a player's (translated) AHL and IHL statistics in a study of how a player's scoring rate changes from ages 21 through 29.*

[* Gabriel Desjardins, "NHL Points-per-Game Peak Age Estimation," *Arctic Ice Hockey* (blog), January 21, 2010, http://www.arcticicehockey.com/2010/1/21/1261318/nhl-points-per-game-peak-age.]

But I can sense you reaching for the dog's leash, so let me jump to the more elegant solution from baseball, the delta method, which involves creating matching pairs of adjacent seasons. Tango describes this process of building an age curve as calculating the average season-to-season difference (delta) in the chosen statistic for all players who competed in at least a certain numbers of games and grouping them by age.*

[* Tom Tango, "How to Be Careful with Aging Curves," *Tango Tiger* (blog), September 5, 2013, http://tango

tiger.com/index.php/site/comments/how-to-be-careful-with-aging-curves.]

For example, Toews scored 0.89 points per game at age 25 and 0.81 points per game at age 26, for a difference of −0.08 points per game. That's called a matched pair. By taking the average of all matched pairs for players going from age 25 to 26, we can get an indication of the average increase or decrease in a player's scoring rate at that age. Of course, scoring can also be measured on a per-minute basis or in terms of percentages or even in absolute terms, but you get the idea.

The whole principle is to isolate age as the only variable in the equation, so that any change in scoring can be attributed *solely* to that one factor. Although a player's statistics can change from one season to the next for any number of reasons, these factors should cancel out over a large enough sample size, leaving age as the only remaining variable. For example, some players will score more because they had better linemates, but a roughly equal number will suffer a downgrade in playing conditions, and it should all average out. We will therefore be left with the assumption that any changes in that group's average statistics were due to their one remaining variable: age. That adjustment can then be applied to the scoring totals of any of today's 25-year-olds to predict how he'll do in the upcoming season.

There is still one particularly notable issue with this assumption that I simply can't skip over, and it's caused by puck luck. As we have already covered, any observed performance, such as Toews scoring 0.89 points per game, is actually a combination of both the player's own skill and some random variation, and that random variation won't completely cancel out.

Since observed performance has an obvious impact on opportunity, that means that both skill *and luck* will determine whether or not a player gets the opportunity to play the following season, and therefore the opportunity to form another matched pair. That will result in an uneven distribution of puck luck in the first half of a matched pair, especially as the sets grow older.

At the risk of getting your dog some exercise, let me explain this a little further. Ideally, we need our collection of matched pairs to include roughly equal numbers of players who had hot and cold seasons, so that their luck will cancel each other out, just like all the other factors, and leave age as the only variable. But, in reality, many unlucky players don't get the opportunity to play the following season, especially the older veterans. Similarly, borderline NHLers who had a lucky season will get the opportunity to keep playing. The net effect is that there are a lot more lucky seasons in the first half of the matched pairs than the second. That will pull down the average season-to-season change at that age. This is a type of bias known as selective sampling.

Let's illustrate this with an example. Ville Leino scored 15 points in 58 games for Buffalo in 2013–14. A portion of his disappointing season was due to a dose of bad luck that saw his team score on just 3.9% of its shots while he was on the ice at even strength. Since those poor results prevented him from landing an NHL job the next season, he disappeared from our list of matched pairs. With league-average luck, Leino may have scored 30 points in 2014–15, which would slightly boost the age curve. The impact doesn't always have to be so dramatic that it knocks a player out of the lineup entirely; an unlucky season usually takes on the more subtle form of pushing a player down the depth chart.

Until statisticians have the power to instruct coaches on whom they should play, the solution is to identify and remove the impact of random variation from a player's statistics prior to forming the matched pair, leaving only the skill component. This is known as regressing toward the mean, as explained in the previous section.

One last adjustment to be mindful of when building age curves is the year-to-year fluctuations in overall league scoring levels. Even a single season can have a big impact on a number of statistics, especially in the wake of a league expansion or a major rule change.

Consider annual fluctuations in save percentage, for example. Returning to Burtch's work on goalie aging curves, he accounted for the variable nature of goaltending statistics by basing his age curve on "the average change in save percentage by standard deviations from the seasonal mean for each age group."* In other words, rather than look at how a goalie's save percentage changed from year to year, Burtch calculated the difference between a goalie's save percentage and the league average and how that changed for each goalie at each age. This allowed him to base his study on a far larger volume of data going all the way to when save percentage was first officially recorded back in 1984, despite the significant fluctuations that occurred throughout the 1980s and 1990s.

[* Steve Burtch, "Theoretical Goaltending Aging Curve," *Pension Plan Puppets* (blog), September 5, 2013, http://www.pensionplanpuppets.com/2013/9/5/4696042/theoretical-goaltending-aging-curve-Bernier-Reimer.]

Now there are a lot of age curves out there—too many, in the eyes of Alan Ryder and his dog—and I have summarized many in this section. Those of you who have already read these many pages on age curves probably can't imagine that there's actually still more ground to cover. Suffice it to say, all age curves that I covered follow the same basic pattern. Most players hit their peak by age 24 or 25 then decline gradually until age 30, at which point their performance can begin to tumble more noticeably, with the risk of absolute collapse by age 34 or 35. All team-building models should take that into account. Now why didn't I just write that up front?

5. INCORPORATING INDIVIDUAL CAP HITS

Converting the chosen metric(s) into dollars isn't nearly as difficult as it sounds, because hockey's 3-1-1 rule states that every three goals scored or prevented results in one point in the standings (and costs $1 million in cap space, but we'll revisit that momentarily). That means that any metric that is recorded in terms of goals, wins, or points can be converted to cap space quite easily.

In fact, that's exactly the basis for the cap-adjusted version of GVT I unveiled way back in 2009, dubbed goals versus salary (GVS).* Since GVT is measured in goals a player scores or prevents relative to a replacement-level player, like an AHL call-up, a player's GVS can be calculated using the following simple formula:

[* Rob Vollman, "Howe and Why: The NHL's Top Values," *Hockey Prospectus* (blog), August 10, 2009, http://www.hockeyprospectus.com/puck/article.php?articleid=240.]

GVS = GVT – (cap hit – league minimum)x3

Essentially, to calculate a player's break-even GVT, subtract the cost of a replacement-level player (the league minimum) from his cap hit and then multiply that total by three goals per million dollars. Then a player's GVS is that individual's actual GVT minus this breakeven point.

Notice that this calculation allows for a negative result, which introduces the key concept of a negative value player, something Sean McIndoe wrote about last season.* Essentially, he uses the term to describe a player that may (or may not) be an otherwise useful and productive member of the lineup but who has a cap hit that is so steep that his overall value to the team is less than his cap-based peer group. This reality will be reflected by a negative GVS, which isn't to suggest that he is a bad player but merely that his contributions came at a higher cost per goal.

[* Sean McIndoe, "In Praise of the NHL's Negative Value Guy," *Grantland* (blog), February 5, 2015, http://grantland.com/the-triangle/in-praise-of-the-nhls-negative-value-guy/.]

While hockey's 3-1-1 rule is a handy way of estimating value on the fly, the team-building model should never include fixed numbers like that. The variables should be recorded with more precision and recalculated every season. Even today, the rising cap space and league minimums have actually made that handy rule a significant overestimate.

How can we objectively figure out the cost per win for ourselves, independent of metrics like GVT? One method, popularized by Gabriel Desjardins in 2011, is to take the total amount of money that is spent on all players over and above the league-mandated minimum and then divide that by the number of points in the standings that all those players earned.*

[* Gabriel Desjardins, "How Much Do Wins Cost?," *Arctic Ice Hockey* (blog), October 12, 2011, http://www.arcticicehockey.com/2011/10/12/2482642/how-much-do-wins-cost.]

The key point here is that a team composed exclusively of replacement-level players would not finish with zero points. After all, even the lowly 2013–14 Buffalo Sabres, who were arguably the worst non-expansion team in several decades, still managed 52 points. Since the league average is 91 points, that means that each team's $60 million in cap space above and beyond what it would take to pay a 23-man roster at the league minimum has to secure about 40 points. That brings the estimates closer to $1.5 million per point—and rising. Unfortunately, hockey's 6-2-3 rule doesn't have quite the same ring to it.

Applying this kind of approach to every season since the cap was introduced back in 2005, it appears that the cost per point has already doubled. Why? Because the cap ceiling has been rising far faster than the cost of a roster full of replacement-level players.

The following table includes every variable in the calculation, including the league salary cap, the league minimum, what it would cost to build a 23-man roster at that league minimum, the amount of cap space left over, and the cost of each additional point in the standings (which is achieved by dividing by 40 points).

THE COST FOR A POINT IN THE STANDINGS, 2005–06 TO 2015–16

SEASON	SALARY CAP	LEAGUE MIN	23-MAN ROSTER	CAP SPACE LEFT	COST/ POINT
2005–06	$39,000,000	$450,000	$10,350,000	$28,650,000	$716,250
2006–07	$44,000,000	$450,000	$10,350,000	$33,650,000	$841,250
2007–08	$50,300,000	$475,000	$10,925,000	$39,325,000	$983,125
2008–09	$56,700,000	$475,000	$10,925,000	$45,775,000	$1,144,375
2009–10	$56,800,000	$500,000	$11,500,000	$45,300,000	$1,132,500
2010–11	$59,400,000	$500,000	$11,500,000	$47,900,000	$1,197,500
2011–12	$64,300,000	$525,000	$12,075,000	$52,225,000	$1,305,625
2012–13	$70,200,000	$525,000	$12,075,000	$58,125,000	$1,453,125
2013–14	$64,300,000	$550,000	$12,650,000	$51,650,000	$1,291,250
2014–15	$69,000,000	$550,000	$12,650,000	$56,350,000	$1,408,750
2015–16	$71,400,000	$575,000	$13,225,000	$58,175,000	$1,454,375

The cost for a point has gone up in nine of the 11 years of the CBA's existence. It may finally be levelling off, especially since the upcoming rise in the league minimum means that a roster of 23 replacement-level players will cost $14,950,000 in 2016–17 and $17,250,000 in 2021–22. Of course, such a team would be in violation of the league's cap floor, but it remains a valid theoretical construct for the purposes of these calculations.

Now that the final digit in hockey's 3-1-1 rule has been explained (and invalidated), let's tackle where the first number comes from. The simplest estimate is to divide all goals by all (regulation time) points. This may vary from season to season, but it's roughly 6,720 goals divided by 2,460 points, which equals 2.73 goals per point.

A more visual explanation of why we believe it takes three extra goals to earn one additional point in the standings comes from Eric Tulsky, in one of his very first contributions to the world of hockey analytics.* He simply built a graph with every NHL team's goal differential on one axis and their points in the standings on the other axis, and he observed that every three goals resulted in approximately one extra point. I've included an updated version of that chart here so we can see that for ourselves.

[* Eric Tulsky, "A Peek Behind the Curtain: How Do Numbers Get Analyzed?," *Broad Street Hockey* (blog), March 1, 2011, http://www.broadstreethockey.com/2011/3/1/2018705/a-peak-behind-the-curtainhow-do-numbers-get-analyzed.]

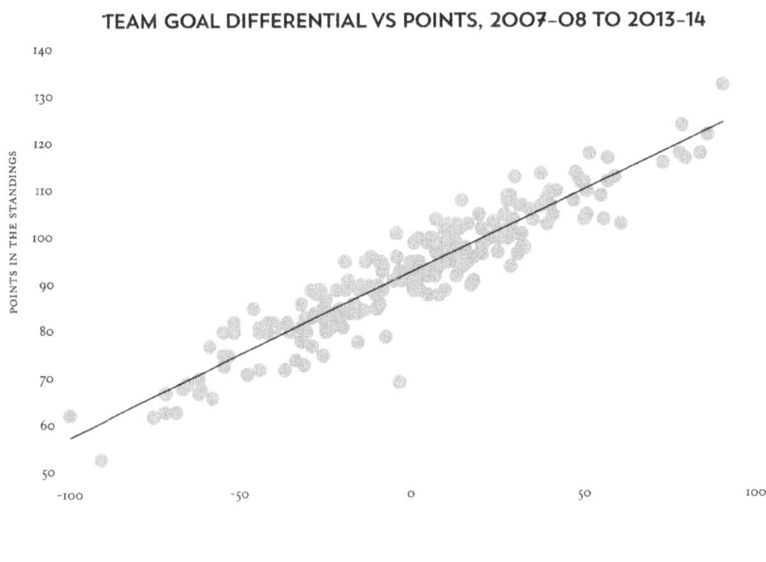

GOAL DIFFERENTIAL

Select a few points along that trend line, and notice that for every additional three goals on the horizontal axis, the line goes up by one point in the standings on the vertical axis. Technically it's closer to 2.86 goals for every point in the standings, and while three is a handy rule of thumb, there's no reason for the cap-building model to be imprecise.

Also, observe how close the relationship between goal differential and points in the standings has been. The only dots that aren't closely hugging the line are generally from the lockout-shortened 2012–13 season, such as Tampa Bay's disappointing 68-point-pace season despite a roughly even goal differential. This reinforces the premise on which this entire model is built: that goals are very closely related to winning.

This chart is also a good visual confirmation of how teams really can't get much worse than Buffalo's recent 52-point season. While all goals are treated equally, technically there is a (slightly) decreasing value at the extremes. That is, a few extra goals for an outrageously good or atrociously bad team won't change their position in the standings as much as those same three goals would help a league-average team. However, that's a phenomenon so obscure that it has no place in this model, which assumes that all goals have the same value.

Incorporating Additional Factors

Having a statistical perspective on a team, player, or situation is only useful if it can be translated into information that leads to real-life applications and decisions. In this case, how can our contract-forecasting model incorporate practical concerns and fit into an overall team-building strategy?

1. INCORPORATING ADDITIONAL INFORMATION

Since this model is based on a high-level, catch-all statistic, it would be advantageous to include a way to incorporate more specific information, both statistically and from elsewhere in the organization. In general, this can be achieved in two ways. First, the results of this team-building model can be provided as additional input into an organization's current decision-making process around trades and free agency. The more intriguing option is to develop a process within this model where a player's value can be manually increased or decreased, either in absolute terms or on a percentage basis, as a result of more refined statistics and/or traditional scouting analysis. This model itself can then be used as the central pillar in the teambuilding strategy.

2. CALCULATING ACQUISITION COSTS

All players have acquisition costs that are not necessarily the same. Some target players are on ELCs or restricted free agency, and they will take additional assets to acquire. This greater cost is frequently higher than the player's total expected value, which is why they tend to stay put.

As for veterans, some of them are tied up in complex, long-term deals, while others are easily available as unrestricted free agents. There are those with personal attachments to certain cities, like Mike Fisher in Nashville, or who have alternative opportunities outside the NHL altogether, like Ilya Kovalchuk and Alexander Radulov in Russia's KHL. On the other side of the spectrum, we occasionally hear about superstars who have to be moved out of town quickly after unfortunate off-ice incidents and may be available for less.

Another complexity is the no-movement clauses many contracts include nowadays, which restrict a player's availability to certain select teams. Not only does that make some players completely impossible to acquire, the limited supply of possible destinations can also reduce their asking price in a trade.

Barring one of the aforementioned situations, the overall acquisition cost of players in a trade should usually be even. Unless there's a large disconnect in an organization's evaluation of a particular player, acquiring a collection of players of a certain value should require a group with the same approximate value in return. Therefore, the primary purposes of trades are to improve the team's mix in one area or another and to build toward a window, not necessarily to improve a team's overall value. Unless you can find a real sucker.

3. BUILDING TOWARD A WINDOW

Building a legitimate contender for the Stanley Cup is presumably the goal for all NHL organizations. There may be temporary exceptions, but this is basically a league exclusively consisting of players and teams who have the ultimate goal of being the best in the world. Consistently icing a playoff-level team is one means to that end, but surfing controlled peaks and valleys is a way to improve the odds slightly more.

Teams do need to build toward windows. Consider a hypothetical team with a total expected performance of 600 goals above replacement level over a five-year period. If that performance is evenly split at 120 goals per season, then this will be a consistently competitive team that might never see the second round. On the other hand, if this hypothetical club is built toward a window, exchanging players who are peaking right now for those who will be more competitive down the road or vice versa, then it will compete for the Stanley Cup at the probable expense of being less competitive in a couple of the other years.

Of course, this approach might not make sense for a general manager who could lose his job if his team struggles. In some markets, there can be monetary concerns at play as well, even though Wesley Chu recently demonstrated that attendance doesn't really change with reduced scoring and/or winning.* Nevertheless, the point remains that sometimes there are external factors that make building toward a window impractical.

[* Wesley Chu, "Winning, Scoring Hardly Matter to Attendance Levels," *Hockey Prospectus* (blog), January 30, 2015, http://www.hockeyprospectus.com/chu-winning-scoring-hardly-matter-to-attendance-levels/.]

4. GETTING THE RIGHT MIX

Blindly assembling a team of players with the best value could result in an imbalance between forwards, defencemen, and goalies, or it could result in too many penalty-killers and not enough power-play specialists or a team completely lacking in some specific area, such as faceoffs or the shootout. While it makes sense to take advantage of market inefficiencies and acquire undervalued players, that can't come at the expense of the team's overall balance. But what *is* that best overall balance?

According to Kevin Allen, "The Los Angeles Kings and Chicago Blackhawks are proving that to stay on top in the NHL, you need a Norris Trophy-calibre defenceman, a superstar two-way centre, a difference-making goaltender and a general manager who knows the secret formula for fitting all of those players comfortably under the salary cap."*

[* Kevin Allen, "Blackhawks, Kings Build for Success in Salary Cap Era," *USA Today,* May 21, 2014, http: //www.usatoday.com/story/sports/nhl/columnist/allen/ 2014/05/21/blackhawks-kings-build-successsalary-cap era/9397115/.]

While most NHL teams could do a lot worse than to simply emulate successful models like these, it's a lot harder than it looks, and things change quickly. Already I would start to question the wisdom of the second-last point, investing $58 million over 10 years or $36 million over six years for 30-year-old netminders like Jonathan Quick and Corey Crawford. Besides, a goal is a goal, so does it really matter if it comes from the centre or a winger or if it's a result of six even defencemen working together versus an elite top pairing?

The right mix is whoever can get the job done, preferably at a discount. The appeal of keeping the team-building model up to date is being among the leaders in figuring out the market's current inefficiencies and tuning the team's balance accordingly. Should the team pursue a high-scoring forward or a shutdown defenceman? Is it preferable to be built around a single franchise player or several elite players or to have a well-balanced roster? Should players be specialized into specific roles or should the team be constructed with do-it-all players with fewer weaknesses?

There is no single right or wrong answer to such questions, or at least none that applies to all teams, in all situations, and at all times. The answer to each of these questions depends on the team's current assets, the players who are available, and the market inefficiencies that currently exist.

The All-Cap Team

Building a team in the salary cap era is not a static exercise, but rather a constant quest to improve one step at a time. Unveiling an all-cap team, an idea borrowed from Ryder's "2011 NHL Review," contradicts that statement.* On a real team, you don't get to cherry-pick your lineup from scratch, and you certainly don't get access to more than a few of the game's top talents.

[* Alan Ryder, "2011 NHL Review," Hockey Analytics, 2011, accessed April 9, 2015, http://www.hockeyanalytics.com/Research_files/2011_NHL_Review.pdf.]

So why am I doing it? First of all, it's fun. You readers just waded through several pages about the current CBA, statistical regression, and how age curves are built, so you certainly deserve a payoff that's a little more amusing. Furthermore, everything you've read so far starts to make a lot more sense when the end result can be seen—however implausible an example it may be. Finally, a look at the best (and worst) active contracts can not only help determine which specific players to pursue and avoid, it can help establish some general rules and guidelines.

What follows is the all-cap team I came up with based on the remaining value of all NHL contracts as they stood at the start of the 2015–16 season. Slightly different results and rankings are possible depending on the metric, age curve, regression factor, projected cap increase, and three-year weighting used in alternative models, but these are essentially the best value contracts in the league right now, not just for this season but into the future.

To remind everyone of where the total contractual value column comes from (value), it is the sum of the player's projected GVS over the duration of his contract. Specifically, I've taken the three-year weighted average of each forward's and goalie's GVT and each defenceman's deltaSOT, regressed it back to the mean to remove the effect of random variation, applied an age curve to predict future performance, and compared it to the value that's expected for the same cap space using the conservative assumption of a 4.7% increase in the league's cap ceiling. Yes, we certainly covered a lot of ground to get to this point. Now here's my dream team.

THE HOCKEY ABSTRACT ALL-CAP TEAM, AS OF AUGUST 2015

PLAYER	TEAM	AGE	CAP HIT	EXPIRES	GVT	VALUE
Carey Price	MTL	28	$6,500,000	2018	18.5	18.2
Andrew Hammond	OTT	27	$1,350,000	2018	5.5	11.3
Marc-Édouard Vlasic	SJS	28	$4,250,000	2018	12.6	15.1
Jake Muzzín	LAK	26	$4,000,000	2020	13.1	30.7
Cam Fowler	ANA	24	$4,000,000	2018	11.6	13.9
Roman Josi	NSH	25	$4,000,000	2020	10.9	20.5
Marco Scandella	MIN	25	$4,000,000	2020	10.7	19.7

PLAYER	TEAM	AGE	CAP HIT	EXPIRES	GVT	VALUE
Ryan Ellis	NSH	25	$2,500,000	2019	9.3	21.6
Tyler Seguin	DAL	24	$5,750,000	2019	19.1	34.7
John Tavares	NYI	25	$5,500,000	2018	16.3	19.1
Gustav Nyquist	DET	26	$4,750,000	2019	11.9	14.3
Max Pacioretty	MTL	27	$4,500,000	2019	16.0	31.5
Kyle Turris	OTT	26	$3,500,000	2018	10.8	14.6
Mark Stone	OTT	23	$3,500,000	2018	13.1	22.4
Tyler Johnson	TBL	25	$3,333,333	2017	16.3	21.2
Ondrej Palat	TBL	24	$3,333,333	2017	14.6	17.7
Tomas Tatar	DET	25	$2,750,000	2017	11.1	13.2
Nick Bonino	PIT	27	$1,900,000	2017	8.9	12.2
David Pastrnak	BOS	19	$925,000	2017	5.9	10.8
Nikita Kucherov	TBL	22	$894,000	2016	13.9	13.2

According to the numbers, this hypothetical team would be staggeringly successful. This roster has a total GVT of 250.1 for the coming season, which is ridiculously high. In fairness, all this talent would mean that most of these players would have to take on lesser roles and that reduced ice time would prevent the team from reaching its full 250-goal potential, but even at around 180 goals (above that of a replacement-level team), the team would still be good enough to win the President's Trophy.

Furthermore, the contract expires are nicely staggered, with only a single one at the end of this season. There are five more the following season, seven more after that, while the remaining seven are signed for at least four years. While I'm sure that a better all-cap team could be theoretically crafted on a year-by-year basis, the whole point of this exercise is to build a team that will be competitive for years to come.

The goaltending selection was easy, and you didn't exactly need to purchase a copy of *Hockey Abstract* to pick Carey Price as the dream starting goalie. For his detractors, Boston's Tuukka Rask or Washington's Braden Holtby would also be good options. For yet another opinion, I re-ran all the numbers using WAR as the catch-all statistic instead of GVT (don't tell Tom), and New Jersey's Cory Schneider and Colorado's Semyon Varlamov bubbled to the top as possible alternatives to Price and Chicago's Scott Darling as an alternative for his backup.

As for the blue line, the conventional wisdom seems to be for teams to build around a single, highly paid Norris-calibre defencemen, like Chicago's Duncan Keith, Los Angeles's Drew Doughty, and Boston's Zdeno Chara. While that makes a great deal of intuitive sense, the safer practice is to select a core of several solid options, all in that $4 million range. There will be more on that in my upcoming rules and guidelines.

I'm obviously quite happy with this collection of defenders, but the WAR-based version of the team-building model isn't as high on Cam Fowler or Marco Scandella, and it's lukewarm on Roman Josi, at best. If you agree with the overall spirit but also disagree with one or two specific selections, both systems agreed on Tampa Bay's Anton Stralman and Anaheim's Sami Vatanen as worthy substitutes.

The presence of just two players on ELCs on the entire roster, Boston's David Pastrnak and Tampa Bay's Nikita Kucherov, suggests that I was trying to "play fair" by choosing a limited number of such players, but that's not it at all. While ELCs are tremendous bargains on a per-year basis, they're capped to just three years or less in length. With the exception of certain blue chippers, it takes at least that long for teams to bring a prospect along and figure out what he can contribute. So while they can be an excellent way to make room for more high-priced stars, you should be careful about how much you rely on ELCs overall.

As for the star forwards, the choices are essentially self-explanatory and probably the least controversial names on the roster. Again, if exception is taken with one or two specific selections, obvious alternatives for a comparable cap hit include Dallas's Jamie Benn and San Jose's Joe Pavelski. St. Louis's Vladimir Tarasenko would be a sensational choice as well, but his $7.5

million cap hit would require a few more ELCs at the bottom of the lineup.

Regardless of how this all-cap team is crafted, it is clearly just a fantasy. No real team can build a lineup like this, which is why it is more helpful to study the best examples of actual NHL franchises heading in the right direction.

Based on the players that repeatedly came up during this exercise, the Tampa Bay Lightning may be the ideal real-world example. The Lightning rank first in the following table, which includes each team's total contractual value, in goals relative to cap space, for its entire combined roster for the current season (as of writing) all the way through the duration of its current deals.

CONTRACT VALUE BY NHL TEAM, 2015-16 THROUGH TO 2022-23

TEAM	2015-16	LONG TERM	TEAM	2015-16	LONG TERM
Tampa Bay	42.9	54.9	Los Angeles	10.2	-37.9
NY Islanders	27.3	46.4	New Jersey	-10.3	-38.8
San Jose	17.4	45.7	Detroit	14.3	-41.4
St. Louis	18.6	16.8	Arizona	-23.8	-46.6
Washington	16.2	13.4	Boston	-3.6	-47.3
Dallas	22.4	11.0	Florida	-14.0	-47.3
Winnipeg	18.2	3.9	NY Rangers	7.3	-47.7
Nashville	16.6	-0.4	Columbus	-12.2	-61.0
Montreal	13.9	-10.1	Carolina	-11.5	-62.8
Minnesota	14.1	-13.7	Chicago	0.9	-63.3
Colorado	-5.0	-15.0	Pittsburgh	5.3	-72.7
Anaheim	31.1	-17.5	Edmonton	-40.8	-77.8
Vancouver	-8.7	-19.9	Philadelphia	-10.2	-87.7
Calgary	5.6	-23.7	Buffalo	-33.6	-96.7
Ottawa	3.9	-25.3	Toronto	-15.7	-108.9

It's not completely fair to compare these teams to one another, given that each of them has varying numbers of players under contract, for varying numbers of seasons. In addition, only a few of them actually spend all the way up to the salary cap, but it does provide a general picture of which teams are in the most flexible position for 2015–16 and down the road.

There's also an important reason why only seven teams are in positive territory over the long term and why all 30 teams don't add up to zero. Essentially, it's because, on the whole, short-term contacts are better value than long-term contracts.

Why is that? Considering it from both sides, very few players would agree to a long-term contract for anything less than their full value, and GMs will offer long-term deals only to the players they feel are most important to the team's future and, therefore, most deserving of a moderate premium. That's why long-term deals don't usually involve discounts.

The remainder of each team's roster is usually composed of short-term value contracts and a steady stream of ELCs, which may change from year to year. All of this means that a team's long-term picture will include only those riskier deals for which it was comfortable paying a premium, while the short-term view will include a bit of everything and even out to zero.

That's why that final column can best be viewed as which teams have the best long-term salary cap flexibility. Tampa Bay, the New York Islanders, and the other teams at the top have very few risky long-term deals and quite a few players on ELCs playing at a discount, giving them the flexibility to take on some higher-priced veterans down the road. In contrast, Toronto, Buffalo, Pittsburgh, and the teams at the bottom have so many long-term commitments at the absolute top dollar that they really don't have much room to manoeuvre and/or improve down the line. Next, let's take a closer look at some of those riskier deals.

The Riskiest Contracts

Sometimes front offices take tremendous risks to acquire or retain either a foundational cornerstone or that one veteran who can boost them up among the legitimate top contenders.

I'm more comfortable calling these risky contracts instead of bad contracts for a few reasons. First of all, these contracts are being graded based on a calculated estimate of how players will perform in the future, not on their actual performance. Technically, we don't know which contracts are bad until after the fact.

Furthermore, I've never really been comfortable with putting a big, fat negative sign in front of someone's value. To reiterate what I wrote earlier, a negative result does not mean that a player doesn't have value or that he's a bad player. Each of these players is contributing just as much as, if not more than, the players on the all-cap team but simply with a cost per goal that's considerably higher.

Finally, I'm a hockey fan, first and foremost. An awareness of which players will require the most inspirational and age-defying elite performances in order for their teams to succeed is one of the reasons I enjoy the game so much. In fact, I fully expect more than a few players on this list to make me look like a real dummy a few years from now.

TOP 30 RISKIEST INDIVIDUAL PLAYER CONTRACTS, AS OF AUGUST 2015

PLAYER	TEAM	AGE	CAP HIT	EXPIRES	VALUE
Dion Phaneuf	TOR	30	$7,000,000	2021	-61.2
Dustin Brown	LAK	31	$5,875,000	2022	-49.0
Shea Weber	NSH	30	$7,857,143	2026	-46.5
Evgeni Malkin	PIT	29	$9,500,000	2022	-45.9
P.K. Subban	MTL	26	$9,000,000	2022	-44.5
David Clarkson	CBJ	31	$5,250,000	2020	-43.0
Henrik Lundqvist	NYR	33	$8,500,000	2021	-42.8
Ryan Kesler	ANA	31	$6,875,000	2022	-41.8
Patrick Kane	CHI	27	$10,500,000	2023	-41.3
Nathan Horton	TOR	30	$5,300,000	2020	-40.9
Jordan Staal	CAR	27	$6,000,000	2023	-40.4
Ryan Suter	MIN	31	$7,538,462	2025	-40.2
Travis Zajac	NJD	30	$5,750,000	2021	-39.7
Phil Kessel	PIT	28	$8,000,000	2022	-36.7
Bobby Ryan	OTT	28	$7,250,000	2022	-36.7
David Krejci	BOS	29	$7,250,000	2021	-36.1
Andrew MacDonald	PHI	29	$5,000,000	2020	-31.9
Zach Bogosian	BUF	25	$5,142,857	2020	-31.7

PLAYER	TEAM	AGE	CAP HIT	EXPIRES	VALUE
Mike Smith	ARI	33	$5,666,667	2019	-30.9
Jason Spezza	DAL	32	$7,500,000	2019	-29.2
Kris Letang	PIT	28	$7,250,000	2022	-27.7
Dan Girardi	NYR	31	$5,500,000	2020	-26.1
Dave Bolland	FLA	29	$5,500,000	2019	-24.7
Jonathan Toews	CHI	27	$10,500,000	2023	-24.4
Jeff Skinner	CAR	23	$5,725,000	2019	-23.9
Ryane Clowe	NJD	33	$4,850,000	2018	-23.4
Brooks Orpik	WSH	35	$5,500,000	2019	-23.1
Pekka Rinne	NSH	33	$7,000,000	2019	-22.3
Claude Giroux	PHI	28	$8,275,000	2022	-21.5
Jimmy Howard	DET	31	$5,291,667	2019	-21.5

Okay, let's call out the elephant in the room. While most of these names aren't unexpected, more than a few of these players are the best in the world and would only be excluded from a team-building model crafted by a complete fool. Let me explain the thinking behind some of these results.

Imagine a team with arguably the league's best goaltender, Henrik Lundqvist, in nets, P.K. Subban and Shea Weber manning the blue line, and high-scoring Russian Evgeni Malkin and Patrick Kane up front. Doesn't that sound like a Stanley Cup contender? If so, then it comes as quite a shock to find all of these names on the list of the riskiest contracts.

The problem is obviously not their talent but their incredible cap hit. This hypothetical team has already consumed $45.3 million in cap space for these five of the league's best players, leaving about $1.45 million per player to fill the remaining 18 positions.

That essentially means that the balance of the roster will be almost exclusively manned by rookies and checking-line veterans.

How competitive would such a team truly be? Their combined GVT of 61 suggests that they'd finish around 32 points above a 51-point replacement-level team, assuming average value from the remainder of its roster, which is outside the playoff picture. And what about four or five seasons from now, when their average age is 35 but they're still consuming that same $45.3 million in cap space? The team will also have been too good to get early draft picks, so it won't even have blue-chip prospects on whom to rely.

And, let's be fair, most of the players on this list aren't fellow superstars but are actually just second-liners (at best) who have been dramatically overpaid.

On the whole, comparing the all-cap team to the list of the riskiest contracts, it appears that the impact of a bad contract is about twice that of a good contract. In other words, it takes two good-value contracts to cancel out one bad one, making the avoidance of bad deals arguably twice as important as the pursuit of good ones. In a sense, analytics may be more useful in raising red flags than in finding undervalued assets, an idea that leads to another one of the upcoming guidelines.

Now that we've built the model and established a basic idea of how to identify good and bad contracts, let's dive into the resulting rules and guidelines

Team-Building Guidelines

Building a team is not a static exercise that starts from scratch. Every team starts with a set of assets, an organizational direction, and a variety of ever-changing player market conditions. Even if there were such a thing as a perfect mix, it would be one that is constantly changing. The ideal approach is for teams to be as efficient as possible with the roster they already have and to always be driving toward a better mix by consistently making the optimal compromises.

While it may not be possible for the model to deliver the *perfect* roster, it can certainly produce a set of guidelines that teams should be pursuing in the given season. To be most effective, these guidelines should be clear and specific, make rational sense, be backed up with objective observations, and be defined in a way that can be tested for validity down the road.

Based on the latest data and the results of the team-building model described in this chapter, I have selected four of the most diverse and illustrative guidelines. Since they are unlikely to remain valid forever, the thinking behind each one is at least as useful as the tips themselves. Let's start with one that was already alluded to in our examination of today's best and worst contracts.

Avoid Bad Contracts

The greatest value of including analytics in team-building decisions isn't to find hidden gems but to avoid truly bad decisions. As Desjardins recounted to Bob McKenzie, "the cost of one really good analytics expert working for a team would be saved with one good decision from that individual—that using analytics for the acquisition of one good player, or not overpaying on one contract extension or another, would be the wisest investment a team could ever make."[4]

The leaderboards earlier in this chapter further demonstrated that a bad contract does a lot more harm to a team's cap situation than a single good contract can overcome. In reality, it could take two or more good-value contracts to negate one truly awful one.

Yes, teams do need to take risks in order to compete, but some of those gambles are simply unnecessary. When our numbers suggested that 31-year-old Brad Richards's nine-year, $60 million contract with the New York Rangers back in 2011 didn't add up, one

---

4 Bob McKenzie, Hockey Confidential: Inside Stories from People Inside the Games (Toronto: Harper Collins, 2014), 89.

anonymous GM told Craig Custance to "Tell the Rangers that so maybe they'll move him. I'll take him."* The GM in question pointed to his intangibles, his clutch play, and the timing of his scoring to explain the huge gap between his calculated value and his steep contract. And yet, Richards was a healthy postseason scratch two years later and was ultimately bought out in 2014. That's the sort of risk that an exclusively traditional mindset could never see, but which was clear as day to the cold, objective eye of hockey analytics.

[* Craig Custance, "GMs Respond to Summer Signings," ESPN, May 1, 2012, accessed April 9, 2015, http://insider.espn.go.com/blog/craig-custance/insider/post?id=405.]

The ability to walk away from a bad deal like that is a good way to rate a front office's team-building strategies, but we rarely hear about the deals that *aren't* made. To continue the example, there were several teams rumoured to be vying for Richards's services in 2011, possibly including Toronto, Los Angeles, Calgary, Tampa Bay, and Buffalo. One or more of these teams may actually have used some objective calculations to walk away when the bidding started to get out of hand.

Helping to expose unnecessary risks is potentially the best application of this team-building model. While a team can always find other players when it misses out on an opportunity, the really bad contracts tend to linger around forever. As we explored earlier, there are only a limited number ways of getting rid of bad contracts, and those options tend to add up to an even higher cost in the long run, acting like a pair of handcuffs binding the team's wrists for potentially several years to come. If Richards's contract had been bought out in the conventional fashion, for example, it would have carried a cap hit of $4.4 million for 12 seasons. That's the sort of hole that could sink a team for a long time.

Reduce Reliance on Free Agency

Free agents are the most expensive players. Since free agents are typically 27 years old or older, teams essentially pay a premium to acquire players who have already peaked and are on the decline, as we discovered while constructing age curves.

That's not to argue that free agents should be avoided entirely, since there are still lots of undervalued players to be found, and sometimes it's the only way for a contender to get a key final piece of the puzzle, like Chicago's Marian Hossa or Marian Gaborik in Los Angeles. That's why the core of a team should be players who were drafted or acquired as prospects, and free agency should be used mainly to acquire complementary players.

There's one other important distinction to make, so let's take a short step back. There are two forms of free agency, unrestricted (UFA) and restricted (RFA).

There's a rather complicated formula to determine which is which, but generally the unrestricted free agents are those who

- are at least 27 or
- started young and have now accrued seven years of NHL experience (of at least 40 games or more, or 30 for a goalie) or
- are unused players who are at least 25 and have three years of experience in any professional league but fewer than 80 games in the NHL (28 for goalies) or
- were previously bought out or
- are unwanted restricted free agents who were not extended qualifying offers; a qualifying offer is based on the player's previous salary and his NHL experience, but it is generally a one-year offer with a small raise if it's under a million.

With the possible exception of the second bullet point, there are rarely bargains among these types of players. Unrestricted free agents are usually paid a little more than their full market value, since a team doesn't need to sacrifice any of its own players, like they do in a trade, or use up a draft pick in compensation to acquire them, as is the case when the free agents' rights are restricted.

The remaining (younger) free agents are considered restricted. That means that other teams may present them with offer sheets, but, once accepted by that player, the team that owns his rights can match that offer sheet and sign the player to that same contract instead. Alternatively, the team may allow the claim and receive compensatory draft picks from the team that made the offer, based on the average salary of this new deal. In essence, it becomes somewhat of a trade.

Due to that acquisition cost, as well as an unspoken code between the league's GMs not to extend offer sheets to their players, restricted free agents tend to cost less cap-wise than unrestricted free agents. There are usually several teams that see an RFA's cap hit as a good bargain, but not enough of one to justify sacrificing draft picks to acquire him. That's why restricted free agents generally offer an average contract value of 1.8 goals, while the average unrestricted free agent actually costs a team about 1.3 goals, for a net difference of 3.1 goals—or roughly the value of a few compensatory draft picks.

To arrive at those figures, and all the others to follow throughout this upcoming examination of free agency, I took all the contracts that were signed from the end of the 2013–14 NHL season until the time of writing, August 2015, disregarding both entry-level contracts and replacement-level contracts, the latter of which I arbitrarily defined as anyone who played in fewer than 70 games in the three seasons used for this projection. Using the player's cap hit, the data that was available at the time of the contract, and the approach described earlier in this chapter, I calculated the average value for the remaining 138 RFAs and 215 UFAs.

The results? In general, there are three key variables when calculating the value of a free agency contract: age, term, and cap hit. If free agency is truly the option a team needs to pursue in order to fill some holes, then the safest contracts are short, low cost, and aimed at younger players. Let's examine each of these factors in closer detail, starting with age.

1. SIGN THEM YOUNG

Oddly enough, older players tend to get a little more money than young players with comparable statistics, generally because their longer track record makes them appear to be safer investments. In reality, younger players are more likely to meet or exceed expectations, especially on longer deals.

The following chart groups all 353 free agents in our study by their age when their deals went into effect, with the expected contractual value computed by our team-building model on the vertical axis. It may look like just a mess of grey dots, but based on the statistically calculated trend line, the value of contracts does decline more rapidly the older the player is when the deal begins.

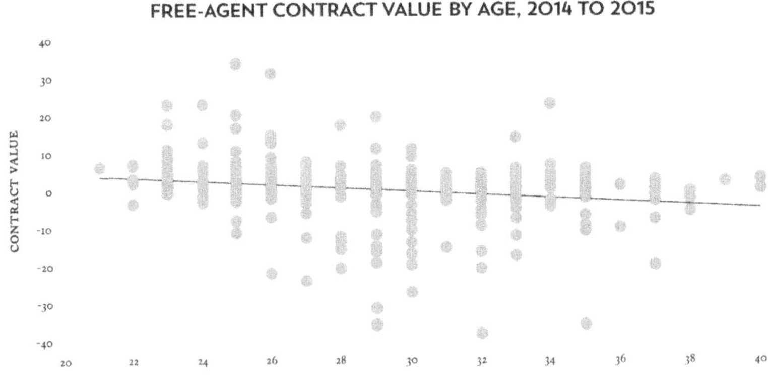

PLAYER AGE

While the overall decrease is barely noticeable to the naked eye, the lack of high-value contracts once players hit their 30s really does stand out. Recently, only two such veterans, 33-year-olds Andy Greene (New Jersey) and Ron Hainsey (Carolina), offer more than a handful of goals of expected value, while there are over a dozen deals that are expected to cost a team at least 10 goals, including most notably Washington's Brooks Orpik.

The main appeal of signing younger players is the relative scarcity of dangerously risky deals, especially before age 27, when they are predominantly still restricted free agents. Edmonton's Justin Schultz, Florida's Dmitry Kulikov, Buffalo's Tyler Ennis, and Montreal's P.K. Subban are essentially the only recent examples of teams taking on any kind of risk with their younger talent, in terms of cap hit, length, or both. In contrast, there are easily a dozen examples of high-value contracts in the same age range, including potential home runs in Nashville's Ryan Ellis,

Los Angeles's Jake Muzzin, and Minnesota's Marco Scandella.

That's why, as far as free agency goes, young and restricted are the ideal contracts. This is partly because restricted free agency adds an acquisition cost and, therefore, lowers competition for a player's services, but also because players in their mid-20s can be expected to sustain their current performance for several years without decline.

2. KEEP IT SHORT

The next important factor is the duration of the deal. In general, the longer the term, the riskier the deal because the team is taking on all the risk if the player gets injured or fails to perform for whatever reason.

On the other hand, if the player signed at a discount, then a long-term deal can turn a good deal into a great one. There are very few low-cost, long-term deals, however—why would any player agree to something like that? In practice, the greater the term, the greater the cap hit, as the following chart illustrates, most notably as contracts approach four years and $4 million per year.

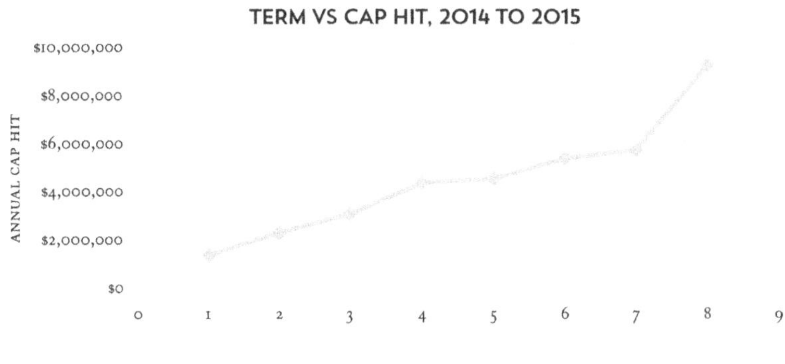

CONTRACT TERM, IN YEARS

This makes perfect sense when you think about it. After all, there is very little incentive for players and teams to agree to deals that are more than three years in length but less than $3 million per season. If the player has any kind of upside potential, then he's going to want more money, even if it means taking on a shorter term to prove himself. And if he really can't develop into anything more than a depth piece, then the team has no incentive to lock him in for longer than three years, at any price.

That's why a long-term contract also means a high-priced contract, almost by definition. According to the following chart of messy grey dots, hitting a home run on deals that run longer than five years is almost completely unexpected. The one outlying exception is St. Louis's 23-year-old Vladimir Tarasenko, who could prove to be a windfall over the entire course of his eight-year deal.

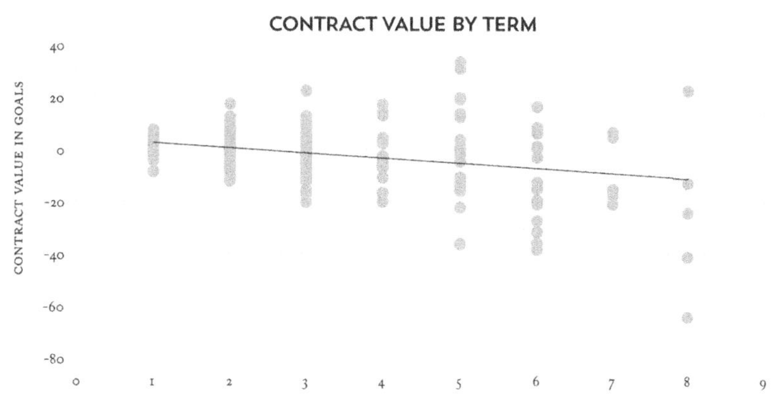

CONTRACT LENGTH IN YEARS

As an additional risk, the longer the term, the more a deal can stray away from the break-even point in either direction. Single-year contracts are all relatively low risk, while deals that extend five or more years into the future can compound those small risks into major gambles.

Of course, all of this is based on my own projections, and I could be proven wrong by the time *Hockey Abstract 2021* rolls out (fingers crossed). I made several choices in building this model, including which player valuation metrics to use, how much to regress the influence of random variation, and the exact nature of the age curve, all of which could have an impact on these results.

Even my assumption that the cap ceiling will increase at a modest 4.7% could affect these results. If the rate of growth in cap space is much higher, then the risk in these long-term deals really goes down. After all, a $6 million contract is a big investment today, but it may turn into a bargain by the 2020–21 season if the cap ceiling quickly explodes to well over $100 million. On the other hand, if revenues fail to grow, then someday Anaheim is going to look awfully foolish using up to 10% of its cap space on a 37-year-old Ryan Kesler or Calgary for an even older Mark Giordano.

With those caveats aside, the value of contracts tends to decrease as the term increases because the performance of older players inevitably starts to decline faster than the cap ceiling grows.

3. KEEP INVESTMENTS SMALL

The strongest statistical correlation between contract value and any of these three factors is with a player's annual cap hit. That makes perfect sense, since it's far easier to achieve value when the cap hit is low, while the damage caused by an underachiever is obviously greatest when the cap hit is high.

Of the three charts full of messy grey dots, the chart showing this trend might be the easiest to interpret with the naked eye. If you use a couple of fingers (or three, if they're particularly slim) to cover the left side of this chart, right up until $2.5 million or so, you can actually see a fairly steep downward trend. It also helps if you squint a little, I think.

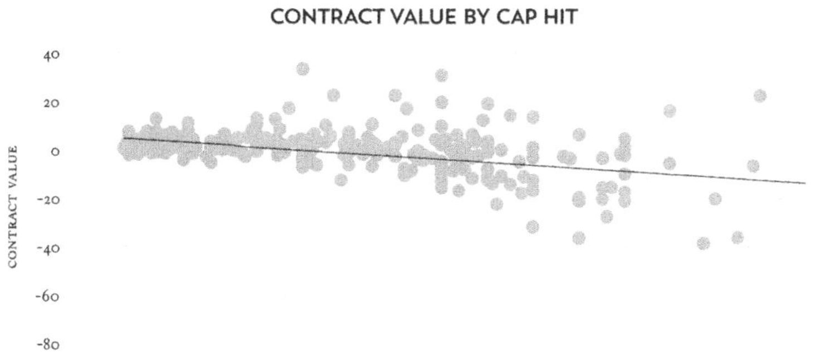

ANNUAL CAP HIT

It's hard to go wrong when signing a free agent to a contract under $2.5 million or so (you can lift your fingers now, by the way). Even particularly questionable investments, like Buffalo's $1.5 million deal with Cody McCormick, the New York Rangers' $1.45 million deal with Tanner Glass, or San Jose's $1.2 million contract with Mike Brown, really don't do their respective teams much harm in the grand scheme of things.

As for the higher-priced contracts, there is still value to be had, especially if the player is young and/or the term is kept short. Even when looking at cap hits in the $4 million range, there are still roughly as many good deals as bad ones. In my view, landing a reliable top-four defenceman in that price range is practically a steal these days, especially if he's young.

We have already observed a strong correlation between term and cap hit. Now let's examine the relationship between age and cap hit (in grey) and age and term (in black), both of which are overlaid on the following chart.

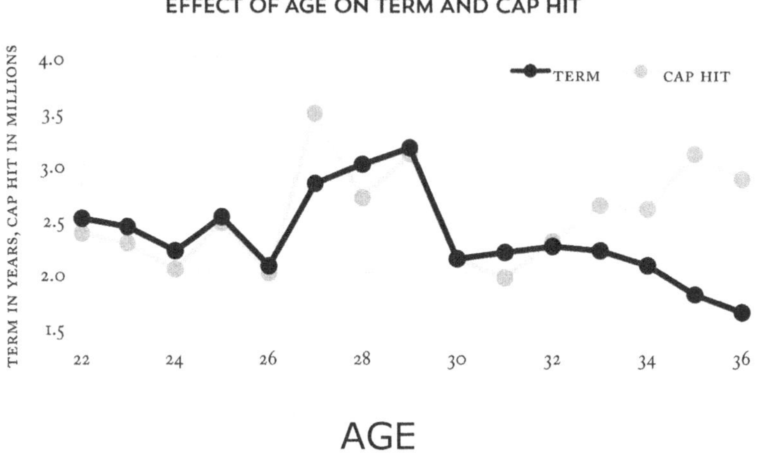

In my view, the unusual shape of both trends in the early 20s is a result of all the rules that apply to restricted free agents, especially the big dip at age 26, right before most free agents become unrestricted.

While it's no surprise that both cap hits and terms are highest for players in their late 20s, the huge plunge at age 30 must be partly psychological. That one birthday shaves a full year off the term and a full million dollars off the table, on average. Better lock down an extension before the candles are blown.

Terms are understandably kept short for players in their 30s, although the cap hit begins to rise. That is likely a result of survivorship bias, since only the NHL's stronger players are going to be offered contracts at that age. In essence, a 34-year-old is going to be offered a large contract or none at all, because there's just no point filling out the depth lines with low-cost veterans instead of young, developing prospects.

4. THERE IS NO HOME-TEAM DISCOUNT

Surprisingly, I found absolutely no home-team discount when I divided the group of free agents between those who signed with new clubs and those who stayed in town. Even when I made the distinction between those who signed extensions and those who re-signed with their current teams *after* becoming free agents on July 1st, there was still no advantage to be found.

Years ago, Desjardins found a 20% discount for players who re-signed with their current teams prior to becoming unrestricted free agents and 40% for restricted free agents,* but I can't reproduce those results today. If anything, the group of re-signed players are signed at a *premium.*

[* Gabriel Desjardins, "NHL: UFA Price per Win in 2010 and Hometown Discounts," *Arctic Ice Hockey* (blog), June 25, 2010, http://www.arcticicehockey.com/2010/6/25/1537033/nhl-ufa-price-per-win-in-2010-and.]

The team-building model calculates each player's value in absolute terms, but converting it to percentages can provide an interesting perspective. To make that conversion, I worked out what every contract's cap hit would have to be for its value to be exactly zero, and then I calculated the percentage difference between that and the player's actual cap hit.

The results? On the whole, unrestricted free agents cost a premium of around 10%, while restricted free agents come at a discount almost as great. However, there is no home-team discount.

Negotiating a contract extension with a pending unrestricted free agent comes at a whopping 25% *premium,* not a 20% discount. Re-signing a player once he has already hit the open market carries only a 5% to 10% premium, which is roughly the same as signing a brand-new player.

Why the higher cost for re-signing pending free agents? Selection bias. Basically, the pool of pending free agents who are re-signed by their current teams and the pool of those who hit the open market are not the same. They're all players their current teams know very well, but the first group is composed of those the teams don't want to risk losing, while the second group is composed of those the teams are perfectly willing to risk on the open market. It makes perfect sense that teams would be prepared to pay a premium for that first group but no more than fair market value for the second.

As for restricted free agents, the relative scarcity of those who actually sign with different teams makes it impossible to repeat this calculation for that group, especially since there's far less of a pressing need to finalize their extensions before the free agency period begins.

5. *SOME MARKETS COST MORE*

Outside incentives are different from city to city, which makes free agency more expensive in some places than in others. Essentially, a player worth $4 million could cost $3.5 million in more favourable markets and $4.5 million in less attractive destinations. Consequently, for some teams, the team-building model should incorporate a market adjustment to account for that additional signing cost or discount.

Why the difference? Certain cities are seen as more preferable to players because of tax advantages, like in Nashville, or perhaps the presence of other stars and/or a shot at the Stanley Cup, as is the case in Chicago. Conversely, ESPN's Craig Custance polled NHL agents last year to find the most commonly listed no-trade destinations and found a few common cities that players wanted to avoid, including Edmonton, Winnipeg, Buffalo, Ottawa, and Toronto.* It stands to reason that free agents are also less likely to sign a deal with such teams and/or would require a premium.

[* Craig Custance, "Most Common No-Trade Destinations," *Craig Custance Blog,* ESPN, February 18, 2015, http://insider.espn.go.com/blog/craig-custance/post?id=4972.]

Rather than rely on those anecdotal examples, I tried to dispel some of the controversy by objectively identifying and measuring any team-by-team discounts or premiums using recent contracts. Unfortunately, there's not nearly enough data, and many of the results are therefore skewed by a couple of outlying deals and an imbalance between restricted and unrestricted free agents. While the following table should be taken with enough salt to bury an elephant, it found an additional expense in only two of the cities Custance singled out, Buffalo and Edmonton.

CONTRACT VALUE OF FREE-AGENT SIGNINGS BY MARKET, 2014 AND 2015

TEAM	SIGNINGS	VALUE	PREMIUM
Carolina	8	42.5	-41%
Nashville	13	66.4	-34%
Pittsburgh	13	44.5	-33%
Winnipeg	13	19.0	-27%
Minnesota	15	44.0	-22%
NY Rangers	16	21.7	-22%
Dallas	9	19.6	-20%
Toronto	19	22.4	-20%
St. Louis	15	26.3	-17%
Montreal	16	-70.9	-15%
Tampa Bay	10	13.5	-15%
Columbus	12	4.7	-14%
New Jersey	14	19.9	-12%
Los Angeles	9	16.0	-12%
Arizona	9	7.6	-11%
Calgary	15	-7.5	-9%
Chicago	8	-69.8	-7%
Ottawa	12	11.3	-5%
Boston	12	-28.9	-4%
NY Islanders	9	-15.0	-2%
Anaheim	11	-34.0	0%
Washington	10	-25.2	7%
Detroit	13	12.5	9%
Philadelphia	14	-41.7	10%
San Jose	11	2.6	13%
Vancouver	10	-13.8	15%
Colorado	12	-27.7	21%
Florida	10	-45.7	28%
Edmonton	11	-27.5	75%
Buffalo	9	-41.5	101%

Surprisingly, the allegedly unattractive destination of Winnipeg actually found itself near the top of the list, enjoying a 27% discount on its free-agent signings. That ranks the Jets fourth behind Carolina, Nashville, and Pittsburgh. The Predators are likely near the top of this list because of a league-low 37.8% tax rate,* and others may be enticed to sign with the Penguins for the opportunity to play with Sidney Crosby, but it's anyone's guess why the Hurricanes rank first. While I'm sure that there actually is a market adjustment that some cities should make, I'm skeptical whether there's any reliable method to estimate it.

[* Tax information obtained from Gavin Group, accessed April 9, 2015, http://gavingroup.ca/personal income-tax-rates-in-nhl-cities/.]

Another detail that jumps out is the occasional disconnect between the overall value of a team's contracts and the average premium/discount for each contract. Chicago and Montreal, for example, are the two lowest teams in the league with regard to the overall value of their contracts, even though the average player signs at a discount of 11% to 15%. Obviously this is the result of a few high-priced, long-term contracts skewing the numbers and negating each team's otherwise impressive collection of value-priced signings.

That's also why the overall team-building model is based on the absolute value of each contract and not represented as a percentage. After all, a team could have 10 good contracts and one bad one, but if the bad one is a massive long-term deal and the good ones are all single-year deals for depth players, then the team's overall cap situation isn't really that strong.

Regardless of how value is ultimately formatted and how accurately the premium in each market can be estimated with such limited data, the point is that various factors can sometimes raise a free agent's signing cost above his actual value or lower an otherwise overpriced asset into a potential bargain. Front offices need to be aware of all the factors that can affect a free agent's contract offer and adjust both their offers and their reliance on free agency accordingly.

6. AIM FOR 20% OR LESS

How much of a team's roster should be acquired through free agency? Obviously that varies based on the team's current assets and the market inefficiencies of the day, but Adam Gretz found that the first nine Stanley Cup champions following the 2005 lockout used only 21.1% of their cap space on free agents, on average. Furthermore, he found that "No Cup winner in the cap era has had more than six players signed as unrestricted free agents in the playoffs" and that most of the exceptions were small, short contracts. Ultimately, Gretz could find only eight free agents over this span, or about one per year, who signed four-year (or more) deals and went on to win a Stanley Cup with their new teams.*

[* Adam Gretz, "How Much Money Do Stanley Cup Winning Teams Spend on Free Agents?," *Pro Ice Hockey* (blog), 2014, http://proicehockey.about.com/o

d/nhlfreeagents/fl/How-much-money-do-Stanley-Cup-winning-teams-spend-on-free-agents.htm.]

The ideal roster construction leans heavily on the types of players with the highest value, which are those whom the organization either drafted and developed itself or who were acquired at a young age. These are the players who are going to maintain and/or improve their performance in the coming years and for whom the various CBA rules artificially reduce the annual cap hits.

Jennifer Lute Costella looked at the roster construction of all NHL teams last year, and it's easy to see the difference between those built primarily around their own draft picks, like Detroit, Ottawa, Chicago, Los Angeles, St. Louis, and the Islanders, and those that aren't, like Philadelphia, Colorado, Vancouver, Toronto, and Minnesota.*

[* Jennifer Lute Costella, "NHL Roster Construction: Drafted, Undrafted & Acquired Players," *The Committed Indian* (blog), December 19, 2014, http://thecommittedindian.com/nhl-roster-constructiondrafted-undrafted-acquired-players/.]

While no team would want to avoid free agency entirely, the only way for an organization to remain competitive is to reduce its reliance on non-cost-effective free agents. Among those free agents who are chosen, the contracts should be kept short, inexpensive, and targeted predominantly at younger players. And remember that there is no such thing as a home-team discount.

Goalies Are Crapshoots

Without question, goalies are the most valuable players on the team, and they can make or break the season all by themselves. But how can the right one be selected?

If a team were to somehow secure the defending Vezina Trophy winner as their starting goalie every single year, they'd actually only have a 0.919 save percentage since the 2005 lockout. That's only a little bit above league average, and it's certainly not worthy of the over $7 million of cap space that these top goalies require.

The problem is that there's a lot of random variation in goaltending statistics, based on the weak season-to-season correlation of save percentages. That's why our model regresses skater statistics by a little under 20% toward the league average, but a three-season weighted average of goalies is regressed by over half. It's consequently very hard to predict which goalies will provide top value in the years to come.

And it's not just the statisticians who struggle to predict goaltending performances, but those who use traditional analysis as well. Prior to the 2014–15 NHL season, Craig Custance interviewed a panel of anonymous front office decision-makers to divide the league's starting goalies into three tiers.* The results at the end of the season, in the table that follows, suggest that this isn't just a blind spot in the world of analytics—nobody can predict goaltending.

[* Craig Custance, "Ranking Goalies into Tiers," *Craig Custance Blog,* ESPN, October 7, 2014, http://insider.espn.go.com/nhl/insider/story/_/id/11653739/anonymous-league-insiders-rank-30-starting-goaltenders-nhl.]

FRONT OFFICE VIEW OF GOALTENDING, 2014–15*

TEAM	CAP HIT	QS%	SV%	ES SV%
Tier 1	$6.95 million	62.8%	0.924	0.933
Tier 2	$5.27 million	57.9%	0.917	0.922
Tier 3	$3.51 million	53.4%	0.914	0.922
Unranked	$1.27 million	52.9%	0.911	0.918

[* Goalie data obtained from the NHL website, accessed April 9, 2015, http://www.nhl.com.]

The NHL insiders did get the top tier right, but it was composed of just Henrik Lundqvist, Carey Price, Jonathan Quick, and Tuukka Rask. Even the guy in your fantasy pool who drafted Dany Heatley last year can get that one right.

Consider that second tier of goaltenders instead. They were paid almost as much as the top tier of goaltenders, but they had an average performance that was virtually identical to the far more affordable third tier. Even the unranked netminders weren't really that far behind, and they're way more affordable.

Basically, given the unpredictable nature of goaltending, unless a team has one of those four elite goalies, it might as well save $4 million in cap space and pick an option from the bargain bin and then invest the additional funds in a top-pairing defenceman or top-line forward instead.

There are very rarely any low-cost skaters in the Art Ross, Hart, or Norris Trophy races, but there's usually at least one low-cost option vying for the Vezina. In just the past few seasons, finalists like Devan Dubnyk, Semyon Varlamov, Ben Bishop, Sergei Bobrovsky, and Jonathan Quick all carried cap hits in the $2 million range. Open up the save percentage leaderboard for any given season, and you'll find way more bargains there than in a list of the NHL's leading scorers. In fairness, a lower-cost option isn't as likely to battle for the Vezina or to finish among the save percentage leaders as one of the more established goalies, but if he does struggle, get hurt, or start declining with age, that low cap hit makes it less of a blow and gives the team more room to manoeuvre.

And there is always a pool of available good low-cost goaltenders from which to choose. While there's obviously a huge gap between a team's top-pairing defenceman or top-line forward and the best readily available replacement-level option, that's not always the case in nets. Take Dan Ellis, for example. In

2014–15, his $900,000 cap hit ranked 60th among goalies, and he is exactly the kind of temporary substitute that a typical NHL team would call upon when hit with injuries. While replacement-level goaltending can be defined in any number of ways, it's reasonable to classify anyone paid less than Ellis as such, excluding those on entry-level contracts. How well did that set of third-stringers perform? The 14 goalies that fit the "below Ellis" definition that year posted a combined 0.917 save percentage with a 2.39 goals-against average in 334 games, which is pretty close to league average. Not bad for under a million dollars.

This isn't to argue that any third-string goalie will post save percentages above the league average, especially since the 2014–15 season was a little unusual in this regard. Matt Cane studied this perspective with more rigour, over a longer period of time, and with a higher cap hit of $1.5 million. Ultimately, he discovered that a 0.911 save percentage is more typical, compared to a league average of 0.914 over that same period.* Most big-name goalies aren't worth investing the extra $5 million in cap space, especially since it's so hard to predict which ones will outperform that league average.

[* Matt Cane, "What's a Goalie Worth These Days? Determining Replacement Level Goaltending," *Puck Plus Plus* (blog), July 10, 2014, http://puckplusplus.com/2014/07/10/whats-a-goalie-worth-these-daysdetermining-replacement-level-goaltending/.]

In fairness, there's a little bit of selection bias at play in this definition of replacement-level goaltending because the better a goalie performs, the more he'll continue to play, even when the regular goalies come back. Most third-string goalies will not post a 0.911 save percentage and will be quickly sent back down to the AHL, but those who do play well will get a disproportionate amount of ice time and pull up that average. That's a classic case of selection bias.

That's why there's no consensus on exactly what a team should expect from a so-called replacement-level goaltender. Over the years, I've seen a number of different analysts calculate those expectations in a variety of different ways. For example, analysts like Scott Reynolds and Bruce Peter have defined replacement-level goaltending based on a cut-off of games played instead of the cap hit.* Their approach actually introduces selection bias in the opposite direction, as, by definition, the goalies that play poorly are the ones most likely to play the fewest games. Indeed, even high-priced, big-name goalies sometimes play poorly enough to lose their jobs and finish the season below that threshold. Consequently, if the approach Cane and I used reaches a rather optimistic conclusion about the quality of replacement-level goaltending, then Reynolds's and Peter's approach will draw the most pessimistic conclusions about the quality of replacement-level goaltending, which is somewhere around 0.904.

[* Scott Reynolds, "Replacement Level Goaltending," Copper & Blue (blog), May 1, 2012, http://www.coppernblue.com/2012/5/1/2992340/replacement-level-goaltending-in-the-nhl; Bruce Peter, "What Should We Expect from NHL Goaltenders?," Eyes on the Prize (blog), May 10, 2012, http://www.habseyesontheprize.com/2012/5/10/3010919/what-should-we-expect-from-nhl-goaltenders.]

Perhaps Gabriel Desjardins is the voice of reason, calculating expectations for a replacement-level goalie to be around 0.908, or just a few points below the unranked goalies in Custance's informal poll of the experts.* This result was based on the goaltending one would expect from a replacement-level team, whether they used big-name goalies or not. Eric Tulsky later revisited this assessment using one of Tango's suggestions from how replacement level is calculated in baseball and also found some merit to that estimate.**

[* Gabriel Desjardins, "What Do NHL GMs Think Is Replacement Level Goaltending?," *Arctic Ice Hockey* (blog), February 24, 2011, http://www.arcticicehockey.com/2011/2/24/2006968/what-do-nhl-gmsthink-is-replacement-level-goaltending.]

[** Eric Tulsky, "NHL Stats: How Good Is a Replacement Level Goaltender?," *Outnumbered* (blog), April 24, 2014, http://www.sbnation.com/nhl/2014/4/24/5620854/nhl-goaltending-replacement-level.]

In the end, regardless of whether $900,000 gets you a goalie with a 0.908 save percentage or something closer to league average, is that still bad enough to warrant paying up to 10 times more for a big name who may or may not perform at a higher level?

None of this is to argue that goalies aren't the most important players on the roster. However, the high availability of decent goaltending, together with the unpredictable nature of their performance, means that goalies are gambles. Sometimes teams have a great read on a particular goalie, place him in the right situation, and the gamble pays off. Other times, the team would have been better off choosing a lower-cost netminder and investing that same cap space to acquire a top-line forward or defenceman, who will more predictably perform far, far better than a low-paid, fourth-line alternative.

Acquire the Right Defencemen

One of the key reasons to have a team-building model is to identify and exploit market efficiencies by finding undervalued types of players. Right now, one of the key opportunities is on the blue line, but only if an organization knows what it's looking for.

Defencemen and forwards are paid roughly the same, but are they both equally replaceable? When studying how teams fared with and without their minute-eating defencemen, Desjardins found that the absence of a typical top-pairing option could cost a team 26 goals over an entire season, far more than the 10 goals when missing a top-line forward.* Since both types of players are essentially paid the same, there's a real opportunity for teams to improve by building out from the blue line.

[* Gabriel Desjardins, "Replacement Level: How Many Wins Do Injuries Cost?," *Behind the Net* (blog), April 17, 2008, http://behindthenet.ca/blog/2008/04/replacement-level-how-many-wins-do.html.]

The main obstacle to exploiting this potential inefficiency is the difficulty in identifying the truly game-changing defencemen. Scoring-based statistics obviously don't help as much in evaluating defencemen as they do with forwards, and traditional defensive measurements haven't proven to be particularly reliable either. Teams that can isolate and measure the performance of defencemen have a huge advantage over the competition right now.

What would I suggest? Well, Domenic Galamini summarized the current thinking of the analytics community when he argued that "possession rates are particularly illustrative of their effectiveness."* That's exactly why I used the shot-based deltaSOT metric for defencemen in this model, instead of the more goal-based GVT metric I used for forwards. Desjardins would also be pleased to hear that the league-wide spread of this shot-based metric is also much wider for defencemen than for forwards, which assigns added value to those who excel.

[* Domenic Galamini, "Reasons Behind the Failure to Accurately Evaluate NHL Defencemen," *Faceoff Circle* (blog), January 16, 2015, http://faceoffcircle.ca/2015/01/16/reasons-behind-failure-accuratelyevaluate-nhl-defensemen/.]

Based on this perspective, the defencemen worth pursuing are those who fit the classic Nicklas Lidstrom model. That is, those who can play 20 minutes a night, take on top-six opponents in both zones, work both special teams, and shut down opposition scoring without taking penalties and while helping to generate an offensive threat of their own. Obviously defencemen who have achieved these qualities at Lidstrom's level are going to be no secret and will deservedly command maximum dollar, but those who are slightly scaled-down versions of this archetype can still be found at the bargain cap hit of around $4 million.

Take the 2014–15 Tampa Bay Lightning blue line, for example. They drafted Victor Hedman second overall in 2009, signed free agent Matthew Carle in 2013, signed free agent Anton Stralman in 2014, acquired Jason Garrison from the Vancouver Canucks that same summer, and then acquired Braydon Coburn from the Philadelphia Flyers at the 2015 trade deadline. All together, that's five defencemen that nicely fit the Lidstrom profile to one extent or another and who carry a combined cap hit of just $23.1 million. You will not build a better blue line for less, especially not one that can take a team all the way to the Stanley Cup Final.

While most of the Lightning defencemen are slightly more defensive-minded, there are also value opportunities of the more puck-moving variety. I'm not referring to the big-name, high-scoring options like Keith Yandle or the ones who require excessive shelter from tough situations, like Justin Schultz, but those who can take on quality opponents, generate good secondary scoring, and drive possession. Many of these defencemen are discounted because they aren't throwing hits and blocking shots, but moving the puck up the ice quickly and posing an offensive threat of one's own is an equally legitimate way to reduce opposition scoring.

Consider the 2014–15 Nashville Predators as a hybrid example of how to build a value blue line. High-priced superstar Shea Weber, who is arguably the closest active player comparable to Lidstrom, may get all the attention, but GM David Poile has built a solid young lineup behind him.

- The value-priced puck movers include 20-year-old blue chipper Seth Jones, who I still can't believe fell to fourth overall in the 2013 draft.
- 24-year-old Ryan Ellis, another good value, was nabbed 11th overall in 2009.
- Solid third-pairing option Mattias Ekholm was drafted three rounds later.

- For the stretch, Poile also reacquired the underrated Cody Franson from the Toronto Maple Leafs at the 2015 trade deadline.
- There's also Weber's top-pairing partner, two-way 24-year-old Roman Josi, who would fit right in on that previously mentioned Tampa Bay blue line.
- Reliable defensive veteran Anton Volchenkov was signed in 2014 for just $1 million and replaced with Barret Jackman the next year for $2 million.

Add it all up, and five of the six defencemen playing behind Weber are in their prime, and all six together have a combined cap hit of around $15 million. Hey, maybe you *can* build a better blue line for less than Tampa Bay's $23.1 million.

While puck-moving defencemen and the "Lidstrom light" variety can often be found at a discount, not everyone is a bargain. In particular, the type of defencemen who are overvalued right now are of the more one-dimensional variety. While that does include the purely offensive-minded defencemen who require shelter and protection in their own end, I am mostly referring to the stay-at-home variety. They do have some value, but those big, hard-hitting, penalty-killing shot-blockers who assist very little offensively or in terms of puck possession tend to come at an unreasonable premium. It never hurts to have one in the lineup, but the readily available and low-cost third pairing options may be the wiser investment.

Finally, markets adapt quickly, and these particular opportunities on the blue line might not exist for very long. There are already several NHL teams that have taken advantage of these inefficiencies, and the pendulum may soon swing in the other direction and place too high a value on these defencemen. The more important takeaway is the process to identify market inefficiencies, which is yet another reason to always keep the team-building model up to date and to stay flexible to the changing conditions.

Closing Thoughts

If I was hired by an NHL organization, the first thing I would do is make a big poster on the wall of what I expected the team's Stanley Cup roster to look like for the season I was building toward. Well, the first thing after I got autographs for my childhood hockey card collection, that is. This roster would be based on the team's current assets as well as who was needed, based on an up-to-date team-building model and the input from the rest of the organization. Given the dynamic nature of this process, many of the specific players would be written in pencil.

This team-building poster would ensure that the assembled roster was strong enough to win the Stanley Cup and would fit comfortably within the available cap space. If kept up to date with changing conditions and player availability, it could help the organization target the right players, to the right contracts, with the right acquisition costs, and toward the chosen window.

The team-building model presented here was not intended to be the perfect model, assuming such a thing even exists, but simply a reasonable example that included all the basic elements that every model must possess.

Specifically, it includes a leading catch-all statistic that can measure a player's value at a high level, which would be calculated relative to the performance of a replacement-level option and relative to the cap space the player required. The player's future performance is projected by taking a multi-year weighted average of his previous performance, calculating and accounting for the effect of random variation, and applying an age curve. In time, this model could be expanded to include more of the granular details of the CBA; more specific detail of each player's acquisition costs; valuations for coaches, trainers, and other key support staff; adjustments based on more detailed statistical analysis, new developments, and/or traditional scouting; and so much more.

Among other uses, this model would supplement the information elsewhere in the organization to evaluate a front office's decisions, manage the team's cap situation, assist in contract negotiations, identify players to be pursued through trades or free agency, and develop a set of guidelines to help steer the team toward Stanley Cup contention one move at a time.

For example, the current data suggests that avoiding bad contracts is more important than finding value-priced deals, that reliance on free agency should be reduced, that goaltending can be a crapshoot, and that there may be some cost-effective opportunities on the blue line right now.

This is the most ambitious use of hockey analytics that I have ever put to print, but it's a project that can be easily completed when each element is broken down and handled in turn. Ultimately, this is one of the best examples of how the field's latest developments can be more than just interesting trivia to include in game coverage and can be directly useful and relevant to the sport.

Obviously no one should ever expect the model to be foolproof, and no one should ever let a spreadsheet make decisions or run the team, but it makes a lot of sense to include this kind of information at an equal level. So much sense, in fact, that some of the NHL's most successful organizations are already doing it.

WHAT DO A PLAYER'S JUNIOR NUMBERS TELL US?

By IAIN FYFFE

On June 26, 2015, Edmonton Oilers GM Peter Chiarelli came to the microphone in the BB&T Center in Sunrise, Florida, to announce that the Oilers were selecting Erie Otters centre Connor McDavid with the first pick of the 2015 NHL Entry Draft. This brought months of anticipation to an end. Edmonton had won the draft lottery, taking the first pick away from the last-place Buffalo Sabres, who ended up with the second pick, and even with a prospect as highly touted as McDavid available, there was some speculation that Edmonton might take a chance on Boston University's Jack Eichel instead, another young player beloved by scouts. Philip Wischmeyer was among a number of writers to suggest that McDavid should not necessarily have been considered a clear-cut favourite in the draft.* In any other year, Eichel would have been an easy number one. So even with a generational talent like McDavid on the board, it was not absolutely clear whether he would be drafted first overall until Chiarelli made the announcement.

[* Philip Wischmeyer, "Could Jack Eichel Dethrone McDavid as the Top Pick?," *Along the Boards* (blog),

June 19, 2015, http://alongtheboards.com/2015/06/could-jack-eichel-dethrone-mcdavid-as-the-top-pick/.]

The NHL's draft lottery was instituted in 1995, in an effort to discourage poor teams from "tanking" in order to improve their chances of drafting a future superstar. But a certain school of thought would suggest that the league's entry draft (and before 1980, amateur draft) has always been something of a lottery. Aside from a few "can't miss" prospects each year, the draft is akin to a crapshoot, where every year some seemingly good prospects ultimately fail to establish an NHL career, while other late picks or even undrafted players break through to have major-league success.

Even some of those "can't miss" players end up missing. Doug Wickenheiser was drafted first overall in 1980 but did not have the career of a number-one pick. Brian Lawton went number one in 1983, as did Alexandre Daigle in 1993, Patrik Stefan in 1999, and Rick DiPietro in 2000—and none of them had a career that approached what was expected of them when they were drafted. And these are far from the only examples; it seems even professional hockey people have some difficulty forecasting the future of young hockey players, which renders the entry draft something of a gamble.

But without scouting, without eyeballs on skaters, all we have left are statistics, and, surely, junior statistics by themselves are not sufficient to project the professional careers of 17-year-old hockey players. There's just too much uncertainty, too many things between now and then that could change for us to make accurate predictions. That is undeniably true to a certain extent, and stats-based projections of junior hockey players are always going to contain a high degree of uncertainty. But the information you can glean from these numbers increases dramatically if you know how to interpret them, and it's not like professional scouting is free from uncertainty. So we're about to enter into a fairly extensive discussion of the interpretation of junior hockey stats, and, as we'll see, there's a great deal of information buried in them—if you know where and how to look. This is not to say, of course, that drafting players based solely on statistics would be a good idea. The best results are achieved by combining scouting results and statistical analysis, as we'll see.

Scope of the Analysis

First we need to establish the scope of what we're studying here. For the purposes of this discussion, we are going to consider only North American junior players and NCAA players under the age of 21. We will consider all junior leagues of a reasonably high calibre, and we will use 2004 as the focus year from which to draw our data to build the model. This means we'll include all of the Canadian major junior leagues (OHL, WHL, and QMJHL), and the best Junior A leagues (BCHL, AJHL, SJHL, MJHL, and OPJHL), plus American Tier I (USHL), Tier II (NAHL), and Tier III (EJHL[5]) leagues. For the NCAA Division I schools playing in Hockey East, the CCHA, ECAC, and WCHA were also considered.[6] Technically, Canadian university hockey is also considered, but since there are so few players of draftable age in the CIAU, it doesn't really add anything to the discussion. It should be noted that we are not using only a single season's worth of data to build our model. We are focusing on the 2003–04 season, but we'll use all of the career junior stats compiled for every player who played junior that season. This gives us a much broader pool of

---

5 The EJHL no longer exists, having been effectively replaced by the USPHL premier division in 2012.

6 In 2013, NCAA hockey underwent a transition, resulting in the Big Ten and NCHC replacing the CCHA and a number of teams shifting from one conference to another.

information than if we were using only a single season's worth of data.

The coverage provided by examining these leagues is very good. Let's look at the 2004 entry draft for example. A total of 291 players were selected that year, and, of these, 87 players were drafted from European clubs, leaving 204 North Americans. Actually, it was 86 Europeans and one Asian, since the Kings drafted Japanese goaltender Yutaka Fukufuji with the 238th pick. The 204 players drafted from North American clubs can be broken down as follows:

- 112 from Canadian major junior (44 WHL, 41 OHL, 27 QMJHL)
- 28 from the NCAA (12 CCHA, six WCHA, five ECAC, five Hockey East)
- 23 from U.S. tiers I, II, and III junior (12 USHL; eight NAHL, including USNTDP; three EJHL)
- 19 from U.S. high schools (eight from Minnesota, eight from Massachusetts, two from New York, one from Connecticut)
- 18 from Canadian Junior A (12 BCHL, two AJHL, two SJHL, one OPJHL, one MAJHL)
- Three from other Canadian junior
- One from other U.S. junior

Of the leagues listed above, the only ones not covered in this analysis are the MAJHL (the Maritime Junior A league), the U.S. high schools, and the "other" junior leagues. The MAJHL and other junior leagues could be included if we thought it was worthwhile; we'd just have to gather the stats for them. The U.S. high schools are a different story. The data that would be required to analyze U.S. high school stats, even if we focused only on the two most important states of Minnesota and Massachusetts, are both widely dispersed and difficult to obtain. Not only do we need the player stats, which are difficult enough to find for all teams, but we would need player biographical information (most importantly date of birth), which schools are unlikely to release for their players. So the analysis would be incomplete, since we need all the data for all the players involved to perform a proper analysis. If you're looking only at players who later played professional or NCAA hockey, you would open yourself up to selection bias, where the results can be skewed by the fact that you are selectively choosing your pool of players rather than including all available players.

The analysis that we're going to do here, and the model that is built upon that analysis (which is called the Projectinator), can also be done for European players, both in junior leagues and senior leagues, but we have only so much space here, so Europeans will have to be saved for a future update. Similarly, we are going to examine only forwards and defencemen in this book; goaltenders are something of a different beast and will wait for a future publication. Going forward, we're going to use the term "junior hockey" in this section, but bear in mind that this is intended to refer to all of the leagues noted previously.

Identifying the Key Numbers

To begin our examination of junior stats, let's have a look at the scoring leaders from the 2003–04 Ontario Hockey League (OHL) season. The OHL is the highest-calibre major junior hockey league, so generally the best young players in this league should have the best shot at going on to successful NHL careers. But our first glance at the leaders table illustrates why junior stats are, on their face, not extremely useful for player projections. Five players recorded 100 or more points that season, and while one of them is Corey Perry, the other four have played a combined total of just 48 regular-season NHL games. Clearly we can't look just at scoring totals to project a junior player's NHL career.

OHL SCORING LEADERS, 2003–04*

RANK	PLAYER	CLUB	GP	G	A	PTS	PTS/ GP	NHL GP	NHL PTS
1	Corey Locke	Ottawa	65	51	67	118	1.82	9	1
2	Corey Perry	London	66	40	73	113	1.71	722	602
3	Martin St. Pierre	Guelph	68	45	65	110	1.62	39	8
4	Eric Himelfarb	Kingston	67	37	70	107	1.60	0	0
5	Daniel Sisca	Sarnia	67	34	66	100	1.49	0	0
6	Rob Hisey	Erie	63	38	58	96	1.52	0	0
7	Mike Richards	Kitchener	58	36	53	89	1.53	710	482
8	Scott Sheppard	London	68	29	59	88	1.29	0	0
9	Patrick O'Sullivan	Mississauga	53	43	39	82	1.55	334	161
10	John Mitchell	Plymouth	65	28	54	82	1.26	412	149
11	Daniel Paille	Guelph	59	37	43	80	1.36	570	172
12	Dylan Hunter	London	64	26	53	79	1.23	0	0
13	Rob Hennigar	Windsor	68	32	47	79	1.16	0	0
14	Joey Tenute	Sarnia	58	22	56	78	1.34	1	0
15	Ryan Ramsay	Plymouth	61	29	48	77	1.26	0	0
16	Matt Ryan	Guelph	68	42	35	77	1.13	12	1
17	Bryan Rodney (D)	Kingston	67	11	65	76	1.13	34	13
18	Rob Schremp	London/ Mississauga	63	30	45	75	1.19	114	54
19	Andre Benoit (D)	Kitchener	65	24	51	75	1.15	179	48
20	Brad Efthimiou	Mississauga	68	22	52	74	1.09	0	0
21	Stefan Ruzicka	Owen Sound	62	34	38	72	1.16	55	17
22	Patrick Jarrett	Owen Sound	67	26	46	72	1.07	0	0
23	James Wisniewski (D)	Plymouth	50	17	53	70	1.40	551	274
24	Wojtech Wolski	Brampton	66	29	41	70	1.06	451	267
25	Petr Kanko	Kitchener	55	26	42	68	1.24	10	1

Note: (D) indicates defencemen.

[* Individual junior and NHL data obtained from Hockey DB, accessed June 19, 2015, http://www.hockeydb.com.]

Corey Locke was no NHL star, so his appearance at the top of the OHL scoring list probably shouldn't be taken as suggestive that he would be. However, Locke has had quite a good professional career, with over a decade in the AHL and German elite league, surpassing a point per game four times in the

American league. But plenty of men have performed well at the AHL level only to fail to produce in the NHL, and Locke's one point in nine major-league games marks him as one of those. But if he was the best scorer in the world's best major junior league (and his scoring totals were even more impressive the previous season), why shouldn't we expect him to continue to excel in the NHL?

Well, there's more information available than what's included in the table, and that's the key to this whole approach. As it happens, there is one absolutely crucial piece of information missing. It is the single most importance consideration when analyzing a player's junior hockey stats, one that, if ignored, skews all of your analysis beyond redemption. That information is simply a player's age.

Any rational analysis of junior hockey stats must start by considering the age of the players involved. Young men mature and develop extremely rapidly between the ages of 15 and 21, with each year, and sometimes even each month, allowing a significant amount of growth as a hockey player. Directly comparing the statistics of a 19-year-old player to those of a 17-year-old is a fool's errand, as the older player has a very significant advantage in terms of his development as a player. And while it may be true that the 19-year-old is a better player at that time, we're trying to project a future NHL career for each of these players, and a 17-year-old has more growth left than the 19-year-old, so that must be considered.

So let's look again at the top 10 scorers from the OHL in 2003–04, but this time we'll order them from youngest to oldest based on their age in years and months as of September 15, 2004, which was the cut-off date for 2004 entry draft eligibility.

TOP TEN OHL SCORERS SORTED BY AGE, 2003–04*

RANK	PLAYER	CLUB	GP	G	A	PTS	AGE (Y, M)	NHL GP	NHL PTS
2	Corey Perry	London	66	40	73	113	19, 3	722	602
7	Mike Richards	Kitchener	58	36	53	89	19, 7	710	482
9	Patrick O'Sullivan	Mississauga	53	43	39	82	19, 7	334	161
10	John Mitchell	Plymouth	65	28	54	82	19, 7	412	149
6	Rob Hisey	Erie	63	38	58	96	19, 11	0	0
1	Corey Locke	Ottawa	65	51	67	118	20, 4	9	1
8	Scott Sheppard	London	68	29	59	88	20, 11	0	0
3	Martin St. Pierre	Guelph	68	45	65	110	21, 1	39	8
5	Daniel Sisca	Sarnia	67	34	66	100	21, 2	0	0
4	Eric Himelfarb	Kingston	67	37	70	107	21, 8	0	0

[* Individual junior and NHL data obtained from Hockey DB, accessed June 19, 2015, http://www.hockeydb.com.]

It's rarely quite so clean and dramatic, but this beautifully illustrates the most important thing you need to consider when interpreting junior hockey stats. More than any other single factor, age matters. An OHL player scoring 100 points in his first entry draft–eligible season is a great prospect; an OHL player scoring 100 points in his final year of junior eligibility is almost no prospect at all. The difference really is that significant.

Another way to look at this is to examine the NHL careers of the top-scoring OHL players who had not yet been through the draft—that is, those who were 17 or younger as of September 15 of the previous year. As you can see in the following table, every one of these players from 1995 to 2004 had a longer NHL career than six of the top-10 scorers in 2003–04, who were all older at the time they delivered that level of performance. With Joe Thornton and Eric Staal, there are some real superstar players included, even though they did not lead their junior league in scoring in their draft years. You don't need to be a league leader in your first draft year in order to be a superstar prospect, at least in the OHL, since you're playing against significantly older players.

OHL SCORING LEADERS BY DRAFT YEAR, 1995 TO 2004*

YEAR	PLAYER	CLUB	GP	G	A	PTS	NHL GP	NHL PTS
1995	Marc Savard	Oshawa	66	43	96	139	807	706
1996	Jon Sim	Sarnia	63	56	45	101	469	139
1997	Joe Thornton	Sault Ste. Marie	59	41	81	122	1,285	1,259
1998	David Legwand	Plymouth	59	54	51	105	1,057	604
1999	Denis Shvidki	Barrie	61	35	59	94	76	25
2000	Raffi Torres	Brampton	68	43	48	91	635	260
2001	Kyle Wellwood	Belleville	68	35	83	118	489	235
2002	Matt Stajan	Belleville	68	33	52	85	774	361
2003	Eric Staal	Peterborough	66	39	59	98	846	742
2004	Wojtach Wolski	Brampton	66	29	41	70	451	267

[* Individual junior and NHL data obtained from Hockey DB, accessed June 19, 2015, http://www.hockeydb.com.]

The NHL results for players one year after their first draft-eligible season are also impressive, but not to the same extent as for the 17-year-olds. The simple reason for this is that the very best 17-yearolds in the OHL do not play in the OHL as 18-year-olds—they're in the NHL. This is another point to remember: older junior players do not only have a developmental advantage over younger players, they also have less competition from their peers since the best of their peer group leave junior hockey sooner. However, this is not a hugely significant factor, since it is only the very best players who go to the NHL directly after being drafted. It's not uncommon for players who develop into all-stars to spend another season in junior after being drafted.

OHL SCORING LEADERS, DRAFT YEAR PLUS ONE, 1995 TO 2004*

YEAR	PLAYER	CLUB	GP	G	A	PTS	NHL GP	NHL PTS
1995	Jeff O'Neill	Guelph	57	43	81	124	821	496
1996	Cameron Mann	Peterborough	66	42	60	102	93	24
1997	Jan Bulis	Barrie	64	42	61	103	552	245
1998	Peter Sarno	Windsor	64	33	88	121	7	1
1999	Norm Milley	Sudbury	68	52	68	120	29	6
2000	Sheldon Keefe	Barrie	66	48	73	121	125	24
2001	Brad Boyes	Erie	59	45	45	90	762	481
2002	Derek Roy	Kitchener	62	43	46	89	738	524
2003	Corey Locke	Ottawa	66	63	88	151	9	1
2004	Corey Perry	London	66	40	73	113	722	602

[* Individual junior and NHL data obtained from Hockey DB, accessed June 19, 2015, http://www.hockeydb.com.]

There are still lots of good 18-year-old players in the OHL each year, as noted above. It's really not until you get to the 19-yearolds that you get a real migration of talent from major junior to the NHL.

OHL SCORING LEADERS, DRAFT YEAR PLUS TWO, 1995 TO 2004*

YEAR	PLAYER	CLUB	GP	G	A	PTS	NHL GP	NHL PTS
1995	David Ling	Kingston	62	61	74	135	93	8
1996	Sean Haggerty	Detroit	66	60	51	111	14	3
1997	Marc Savard	Oshawa	64	43	87	130	807	706
1998	Jeremy Adduono	Sudbury	66	37	69	106	0	0
1999	Peter Sarno	Sarnia	68	37	93	130	7	1
2000	Norm Milley	Sudbury	68	52	60	112	29	6
2001	Branko Radivojevic	Belleville	61	34	70	104	393	120
2002	Nathan Robinson	Belleville	67	47	63	110	7	0
2003	Matt Foy	Ottawa	68	61	71	132	56	13
2004	Corey Locke	Ottawa	65	51	67	118	9	1

[* Individual junior and NHL data obtained from Hockey DB, accessed June 19, 2015, http://www.hockeydb.com.]

At this point, there are many lesser prospects still in the OHL and sometimes-good prospects like Marc Savard who have NHL talent but have to fight a rep of being "too small for the NHL." We'll discuss the size issue later on when we get into the model itself.

OHL SCORING LEADERS, DRAFT YEAR PLUS THREE, 1995 TO 2004*

YEAR	PLAYER	CLUB	GP	G	A	PTS	NHL GP	NHL PTS
1995	Darryl Lafrance	Oshawa	57	55	67	122	0	0
1996	Aaron Brand	Sarnia	66	46	73	119	0	0
1997	Sean Venedam	Sudbury	66	36	55	91	0	0
1998	Colin Chaulk	Kingston	60	34	62	96	0	0
1999	Dan Snyder	Owen Sound	64	27	67	94	49	16
2000	Jonathan Schill	Kingston	65	39	48	87	0	0
2001	Randy Rowe	Belleville	63	64	38	102	0	0
2002	Mike Renzi	Belleville	68	44	64	108	0	0
2003	Chad LaRose	Plymouth	67	61	56	117	508	180
2004	Martin St. Pierre	Guelph	68	45	65	110	39	8

[* Individual junior and NHL data obtained from Hockey DB, accessed June 19, 2015, http://www.hockeydb.com.]

You don't need a degree in statistics to detect the pattern here. It's abundantly clear that the older a player is when having an outstanding junior season, the less likely he is to translate that into success at the NHL level. The leading scorers in their draft years go on to play far more NHL games and record far more NHL points than any other group of players, and the results show a very distinct pattern, as we can see in the table below.

AVERAGE NHL GAMES AND POINTS FOR OHL SCORING LEADERS BY AGE

PLAYER'S AGE	AVG NHL GP	AVG NHL PTS
Draft year	689	460
Draft year plus one	386	240
Draft year plus two	141	86
Draft year plus three	60	20

You could address this issue by drafting only 17-year-old players, ignoring anyone who has already been eligible for the draft but was not drafted. You'd probably do okay with that approach, since you would still be considering most of the best prospects each year. But it's less than an optimal idea, since you would be omitting a not-insubstantial pool of potential puck pros, and there's no reason to hamstring yourself when systematic analysis can account for this issue without unnecessarily restricting your pool of prospects. Fortunately, with a dollop of relatively simple analysis and some statistical manipulation, you can keep the entire pool of players while not being fooled by the performances of older players.

Adjusting the Raw Statistics

The first thing we need to do when comparing players of differing ages is to put their scoring stats on something of an equal footing. An older junior player will be generally more physically mature and have more hockey sense than a younger player, and their coaches know them to be the superior players and therefore tend to give them better ice time and better scoring opportunities. So the advantage of age in junior hockey is ubiquitous.

1. SELECTING THE DATA

First we need to clarify which numbers we're going to be adjusting for age. If we're building a model based on a player's past performance, you should use as much information as you can, so we're going to be using not only a player's regular-season numbers but his playoff stats as well. And we'll include a player's performance in the World Junior Championship as well.* More games played means more information. But we also need to recognize that both the playoffs and international competition are at a higher level of play than the regular season, so we add 20% to a player's playoff goals and 50% to his World Junior goals, doing the same for assists. Please note that we're not valuing playoff performance more than regular-season performance because the playoffs are somehow more important than the regular season; it's simply an acknowledgement that the average quality of opponent is higher in the postseason because the worst teams in the league do not make the playoffs.

[* World Junior Championship data obtained from the Society of International Hockey Research (SIHR), accessed June 19, 2015, http://www.sihrhockey.org.]

2. NORMALIZING FOR SCORING LEVELS

The player's goals and assists are then prorated to 70 games played and a league goals-per-game of 3.50, to represent a typical Canadian junior league. These are both arbitrary, and the actual numbers used don't matter so long as they're the same for everyone.

3. ADJUSTING FOR LEAGUE QUALITY

Finally, the goals and assists are multiplied by a league quality factor.[7] This recognizes the fact that the quality of play is not constant from league to league. Scoring a goal in the OHL is a more difficult task than scoring one in the QMJHL, and it's more difficult than scoring two goals in the Alberta Junior A league. The table below summarizes the factors used by the Projectinator to adjust for league quality.

PROJECTINATOR LEAGUE QUALITY FACTORS

ORGANIZATION	LEAGUE	FACTOR
CANADIAN MAJOR JUNIOR	Ontario Hockey League	1.10
	Western Hockey League	1.05
	Quebec Major Junior Hockey League	0.95
CANADIAN JUNIOR A	British Columbia Hockey League	0.75
	Alberta Junior Hockey League	0.50
	Ontario Provincial Junior Hockey League	0.50
	Saskatchewan Junior Hockey League	0.40
	Manitoba Junior Hockey League	0.35
NCAA	Western Collegiate Hockey Association	1.45
	Central Collegiate Hockey Association	1.35
	Hockey East	1.30
	Eastern Collegiate Athletic Conference	1.15
U.S. JUNIOR	United States Hockey League	0.90
	North American Hockey League	0.70
	Eastern Junior Hockey League	0.40

7 Rob Vollman, "Translating Data from Other Leagues," in Hockey Abstract (author, 2013), 159–182.

4. ACCOUNTING FOR SAMPLE SIZE

With the quality adjustment made, we have put every player's scoring stats on a roughly equal footing, as if each player played in the same hypothetical league together. But since we started with goals per game and assists per game and then multiplied those by an arbitrary number of games played, we can run into a small sample size problem with some players. A player who played 10 games but scored like the devil in that time will be rated very highly, even though, with that small number of games played, you can't tell how much of that scoring was skill and how much was luck. To compensate for this, the Projectinator uses a rough regression adjustment, where for each actual game played under 100, goals and assists are regressed to the mean by 1%. The mean in this case is the average number of goals and assists recorded by a player of that position and age (in years). This ensures that players playing exceptionally well for a small number of games don't receive too much credit for that in the final analysis.

5. ADJUSTING FOR AGE

At this point, we have player stats that are on equal footing and that represent a good number of games played, giving some assurance that we won't be deceived by a fluky performance over a small period of time. We're finally ready to make an adjustment for the most important factor of all: age. The older a player is, the less impressive his scoring stats are, on a relative basis. Recording a season of 100 points may be impressive for a 17-year-old, but a 20-year-old player doing the same has done nothing terribly special.

Age in the Projectinator is calculated based on one year before the cut-off date for the draft in question. For example, a player born on September 15, 1986, would be considered 17 years and zero months old with respect to the 2004 NHL Entry Draft. This is the youngest age a player is eligible for an entry draft. A player born on September 16, 1985, would also be first eligible for the draft in 2004, since he was born one day too late for the 2003 draft. He would be considered 17 years and 11 months old for the purposes of the Projectinator with respect to the 2004 draft. The table below summarizes the age factor that is applied to scoring stats in order to account for a player's age. These values could probably be refined for better accuracy, but this is what we're using at the moment.

POINT MULTIPLIERS BY AGE IN YEARS AND MONTHS

	0	1	2	3	4	5	6	7	8	9	10	11
15	1.436	1.420	1.404	1.389	1.374	1.359	1.344	1.330	1.316	1.302	1.288	1.275
16	1.262	1.249	1.236	1.223	1.211	1.198	1.186	1.175	1.163	1.151	1.140	1.129
17	1.000	0.995	0.990	0.986	0.981	0.976	0.971	0.967	0.962	0.958	0.953	0.949
18	0.840	0.832	0.824	0.817	0.809	0.802	0.795	0.788	0.781	0.774	0.767	0.760
19	0.674	0.665	0.657	0.648	0.640	0.632	0.624	0.616	0.608	0.600	0.593	0.585
20	0.520	0.512	0.503	0.495	0.487	0.479	0.471	0.464	0.456	0.449	0.442	0.435

The inverse of these multipliers can provide a useful rule of thumb when comparing two players from the same league in the same season. Let's say you have a player just eligible for the draft for the first time, and he recorded 100 points. How many points would an older player need to record to be as impressive as this performance? Well, if he was one year older, he'd need 119 points (100 divided by 0.840), and if he's two years older he'd need 148 points (100 divided by 0.674). The following chart visually demonstrates these relationships, and you can use it when making rough-and-ready comparisons between players.

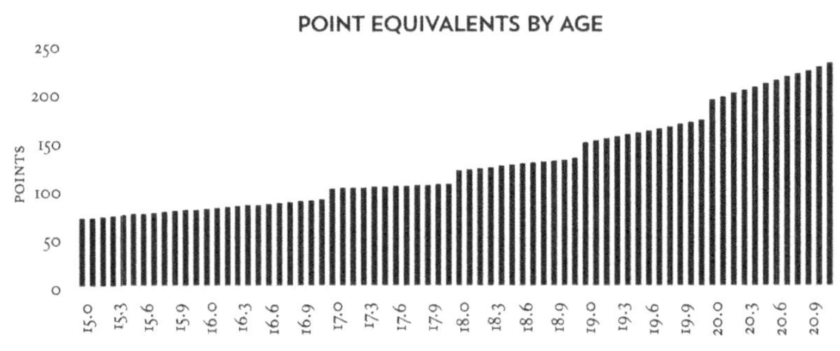

AGE, IN YEARS AND MONTHS (Y.M)

As an illustration of these adjustments, you can look at the career of players who played in junior hockey for several years and see how their apparently increasingly impressive scoring stats as they age are actually an artifact of the advantages that older players have over younger ones in junior. Take Rodney Bauman for example, a winger in the OHL. He was born on March 16, 1983, and so was first eligible for the entry draft in 2001. That season he recorded only 18 points in 47 games and was undrafted. He never was drafted, but as he continued to play in the OHL his production increased steadily, and by the 2003–04 season he put up almost a point per game. But if we look at these points translated for age, we see that his progression was exactly what you would expect for a player getting older and getting more and more advantages over the younger players he plays against.

RODNEY BAUMAN JUNIOR SCORING, TRANSLATED FOR AGE

SEASON	GP	PTS	PTS/70 GP	EQ PTS
2001	47	18	27	35
2002	66	29	31	27
2003	59	36	43	33
2004	49	45	64	33

Many players fit this pattern, which is of course what we would expect because the age factors we use were developed based on the results of many players. But this illustrates that a player who is apparently adding 30% or even 50% to his production year over year may not really be improving his future prospects as a hockey player; rather, he's just progressing as usual for his age.

Now that we have the big stuff out of the way, let's talk about the stuff of being big, for lack of a better way of putting it. We're going to talk about size now.

1998: The Entry Draft of Shame

Probably nothing illustrates the weaknesses in purely opinion-based scouting better than the obsession with size that NHL scouts have often succumbed to. If you begin your scouting process with the idea that smaller players are not as good at hockey as larger ones, your draft rankings will obviously reflect this bias, often in excess of any other consideration. And while there is some reason to believe that size is a general advantage in the NHL (though even this is cloudy because it may simply be the result of a *belief* among NHL coaches that size is an advantage), the degree to which it is true has undoubtedly been overestimated by those responsible for drafting NHL players. And while this bias has seemingly been reduced in recent seasons, it's still worth addressing to illustrate why draft rankings really need an objective component to them, to avoid the sort of groupthink that led to the 1998 NHL Entry Draft results.

It's unlikely that many teams use purely opinion-based drafting anymore, but you also have to make sure that you're using the appropriate objective analysis. The use of measurable data has certainly increased, for example at the annual NHL Scouting Combine, but in that case the things that are measured are purely fitness-related and often have very dubious connections with hockey. While endurance and power output on a bicycle have their obvious informativeness, testing a hockey player's vertical leap, long jump, grip strength, and speed in a shuttle run would seem to be of limited use. It's quite possible that these things are measured because they can be measured, not because they are particularly relevant to drafting hockey players. That is, they decided to measure a bunch of things they could measure rather than starting with an analysis of what they *should* be measuring: what would be most informative about young hockey players. Objective measures are meaningless unless they're relevant to the subject, so not only do you need to be systematic, you need to make sure your system is well designed.

So while a player's size is both objectively measurable and possibly informative, there's good reason to believe that hockey insiders overvalue size in hockey players, sometimes to an extreme degree. The 1998 draft revealed an almost obscene preoccupation with big young hockey players at the expense of good young hockey players. This is not to say that size is irrelevant in the NHL, but perhaps it's just not as important as it may have been assumed to be, at least sometimes, in the past. Especially on defence, it does make sense that size is something of a boon, providing greater reach and greater mass with which to protect and obtain the puck. Determining how much of an advantage it actually is is a challenge, however. NHL teams generally prefer larger players, but it's unclear whether this is because size actually provides a substantial advantage or whether it's because NHL GMs and coaches simply *believe* that size is such an advantage. A bit of both seems likely, but we can't really be sure. The fact that the people in charge of NHL rosters seem to generally believe that size is an advantage makes it very difficult to extract the truth from the results.

How bad was the 1998 draft? Well, to begin with, all of the following players were drafted before Brad Richards, who went 64th overall but who (spoiler alert) is ultimately ranked by the Projectinator as the third-best prospect available in the draft, behind Vincent Lecavalier and David Legwand, who were both rated by the Central Scouting Service (CSS) and actually drafted first and second. Richards has over 1,000 NHL games, nearly 1,000 points, two Stanley Cup championships, and a Conn Smythe Trophy. It was also predictable that he was going to be a very good player; not perfectly predictable, but so much so that waiting until the third round to draft him was irrational, especially when you consider many of the players who went before him, as demonstrated in the following table.

BIG, PHYSICAL PLAYERS DRAFTED BEFORE BRAD RICHARDS IN 1998*

DRAFTED	PLAYER	POS	HEIGHT	LEAGUE	GP	G	A	PTS	PIM
4	Bryan Allen	D	6'5"	OHL	48	6	13	19	126
9	Mike Rupp	F	6'5"	OHL	64	16	11	27	117
20	Scott Parker	F	6'5"	WHL	71	30	22	52	243
21	Mathieu Biron	D	6'6"	QMJHL	59	8	28	36	60
25	Jiri Fischer	D	6'5"	QMJHL	70	3	19	22	112
30	Kyle Rossiter	D	6'2"	WHL	61	6	16	22	190
32	Stephen Peat	D	6'2"	WHL	68	3	14	17	161
34	Andrew Peters	F	6'4"	OHL	60	11	7	18	220
39	John Erksine	D	6'4"	OHL	55	0	9	9	205
41	Maxim Linnik	D	6'4"	Ont Jr B	35	4	14	18	130
42	Jason Beckett	D	6'3"	WHL	71	1	11	12	241

DRAFTED	PLAYER	POS	HEIGHT	LEAGUE	GP	G	A	PTS	PIM
51	Ian Forbes	D	6'6"	OHL	61	2	3	5	164
62	Paul Manning	D	6'3"	WCHA	30	1	5	6	16
63	Lance Ward	D	6'3"	WHL	71	8	25	33	233

[* Individual junior and NHL data obtained from Hockey DB, accessed June 19, 2015, http://www.hockeydb.com.]

Scott Parker's numbers might seem pretty good, but in his case you have to remember the most important lesson contained in the study of junior hockey players: the effect of age. Parker was 20 years old at the time of the 1998 entry draft, and he had been drafted before (but unsigned) in 1996. In his first draft-eligible season, Parker recorded three goals and seven points in 64 games. It seems the Avalanche were not aware of the importance of a player's age when they grabbed Parker rather than Richards, or Simon Gagné or Scott Gomez or Mike Ribeiro or others who were still available and were obviously better bets. But everyone was drafting big players at the time, hoping for the next great "power forward" (ick, a basketball term), apparently neglecting the fact that "big" cannot replace "talented" when it comes to hockey. The same point applies to Lance Ward, who was also drafted in 1996 with scoring stats half as good as his 1998 results, which is to be expected.

All of these players had one thing in common: they were big and physical. This was essentially the entirety of their scouting reports, or at least it was the only part that both the CSS and NHL general managers seemed interested in. The trend continued well into the draft.

MORE BIG, PHYSICAL PLAYERS DRAFTED IN 1998*

DRAFTED	PLAYER	POS	HEIGHT	LEAGUE	GP	G	A	PTS	PIM
67	Alex Henry	D	6'5"	OHL	62	5	9	14	97
81	Justin Morrison	F	6'3"	WCHA	42	4	9	13	8
83	Matt Walker	D	6'4"	WHL	64	2	13	15	124
88	Kent Sauer	D	6'3"	USHL	54	4	19	23	99
92	Eric Beaudoin	F	6'5"	OHL	62	9	13	22	43
95	Andy Burnham	F	6'5"	OHL	38	2	4	6	78
98	Rob Davison	D	6'3"	OHL	59	0	11	11	200
103	Kip Brennan	F	6'4"	OHL	48	0	10	10	188
107	Chris Corrinet	F	6'3"	ECAC	31	3	6	9	22
115	Jay Leach	D	6'4"	HE	32	0	8	8	29
118	Mike Siklenka	F	6'4"	AJHL	54	10	17	27	120
121	Curtis Rich	D	6'4"	WHL	70	3	12	15	204
122	Pat Leahy	F	6'3"	CCHA	28	0	1	1	24
129	Robert Schnabel	D	6'6"	WHL	61	1	22	23	143
130	Gavin McLeod	D	6'5"	WHL	70	3	17	20	98
140	Rick Bertran	D	6'3"	OHL	56	0	9	9	149
141	K.C. Timmons	F	6'3"	WHL	72	11	7	18	139
143	Ryan Flinn	F	6'5"	QMJHL	59	4	12	16	217
148	Chris Ovington	D	6'4"	WHL	68	2	13	15	72
155	Kevin Clauson	D	6'4"	CCHA	36	1	1	2	56
157	Brad Voth	F	6'5"	WHL	70	8	5	13	244
159	Trevor Ettinger	D	6'5"	QMJHL	50	1	2	3	181
172	Jacques Larivière	F	6'3"	QMJHL	68	3	1	4	249

[* Individual junior and NHL data obtained from Hockey DB, accessed June 19, 2015, http://www.hockeydb.com.]

And so on and so on, into the later rounds of the draft. Player after player with little on his resumé except size and physicality. And in case you haven't noticed, none of these players went on to have a significant NHL career. A few of them ended up as enforcers, and even if you put value on such a thing,[8] you have to realize that enforcers are a freely available commodity, and as such you should not be spending valuable draft resources on them. It's likely that NHL insiders learned some lessons from this draft because you simply don't see this size obsession in more recent drafts. Larger players are still generally preferred over smaller ones, but not to the same (let's face it) *insane* degree that they were in 1998.

But even this less extreme preference for size presents a problem to us, because in order to build a model for projecting NHL careers, we need something to build toward, something to value past NHL players in order to forecast what future NHLers might do. In the case of the Projectinator, that something is goals versus threshold (GVT), and GVT is affected by coaching beliefs about player value. You can't accumulate GVT without ice time, and coaches control both the quantity and quality of ice time in the NHL. Even if we used a GVT rate per minute rather than total GVT, this influence would still be felt, due to certain players being given the prime ice time, where GVT can be accumulated more quickly.

---

8 There's no reason to assign value to enforcers, as we explored in Hockey Abstract 2014. See Iain Fyffe, "What Value Do Enforcers Have in Today's NHL?," in Hockey Abstract 2014, by Rob Vollman (author, 2014), 272–286.

So in developing the Projectinator, we could assume that insiders are completely wrong about the importance of size for hockey players, but that would be premature. While we could be fairly confident that the value of size is often overestimated, deciding that it is completely unimportant is indefensible given our current understanding. It would be petulant, arrogant even, to declare that the insiders are completely wrong here. There's no evidence that size doesn't matter, and while here at Hockey Abstract we understand as well as anyone that hockey's perceived wisdom is sometimes entirely the former and none of the latter, you should still have a good reason before you start discarding a particular bit of wisdom entirely.

As such, while size cannot be said to be of preeminent importance, we can't just ignore it either. The Projectinator, therefore, incorporates a small adjustment for offence and a significantly larger one for defence based on the player's size. So to get back to junior scoring stats, we'll apply a small penalty to forwards who are less than 6-foot-1 and to defencemen who are under 6-foot-3.

Creating the Projection

At this point, we've taken a player's junior scoring stats and considered the context of the player's age and league quality, and we've made adjustments for his size and the number of games he's played. The following table summarizes the leaders in these equivalent points (eq pts) for every drafted player who played in junior hockey in the 2003–04 season, including those who were drafted in previous seasons and those not yet eligible to be drafted. In the table, draft year is the first year in which the player was eligible to be drafted.

LEADERS IN JUNIOR EQUIVALENT POINTS, 2003–04

PLAYER	LEAGUE	DRAFT Y	EQ G	EQ A	EQ PTS
Sidney Crosby	QMJHL	2005	58	83	141
Corey Perry	OHL	2003	37	67	104
Tyler Redenbach	WHL	2003	28	61	89
Rob Schremp*	OHL	2004	35	47	82
Andrew Ladd*	WHL	2004	31	51	82
Brandon Dubinsky*	WHL	2004	30	51	81
Gilbert Brulé	WHL	2005	34	47	81
Ryan Getzlaf	WHL	2003	34	46	80
Ryan Callahan*	OHL	2004	43	36	79
Bryan Little	OHL	2005	44	35	79
Clarke MacArthur	WHL	2003	36	43	79
Patrick O'Sullivan	OHL	2003	42	37	79
Dany Roussin	QMJHL	2003	37	42	79
Eric Fehr	WHL	2003	47	30	77
Dave Bolland*	OHL	2004	37	39	76
Rob Hisey*	OHL	2003	30	46	76
Stefan Meyer	WHL	2003	34	42	76
Wojtech Wolski*	OHL	2004	33	42	75
Mike Richards	OHL	2003	31	44	75
Cam Barker (D)*	WHL	2004	23	51	74
Dylan Hunter*	OHL	2004	25	49	74
Chad Klassen*	WHL	2003	29	44	73
Corey Locke	OHL	2003	33	40	73
John Mitchell	OHL	2003	27	46	73
Zach Parise	WCHA	2003	30	40	70
Peter Tsimikalis*	OHL	2004	21	48	69
Alexandre Picard*	QMJHL	2004	34	34	68
Paul Stastny	USHL	2004	26	41	67
T.J. Hensick	CCHA	2004	21	46	67
Kyle Brodziak	WHL	2003	29	38	67

*Player was eligible and available in 2004.

To produce a projection, we need to convert these translated scoring figures into the basis that we use to evaluate hockey players, which is GVT. Specifically, we're trying to project a value for GVT, which is the mean of an NHL player's career average GVT to the age of 27 (which is when a drafted player can normally be controlled by the drafting team) and the average GVT in his two best seasons. This strikes a balance between a player's career value and his peak value.

Creating the Offensive Projection

Converting the translated points into a GVT projection is a relatively simple matter.

1. You subtract a baseline number of points (15 for forwards, five for defencemen).
2. Then divide the result by six for forwards and seven for defencemen.
3. A small size adjustment is then applied, where players above 5-foot-11 receive a small increase and those below a decrease, based on how far above or below they are.
4. Players who played in the World Junior Championship also get a boost, in recognition of the scouting efforts of the professionals.
5. Finally, a threshold value is subtracted (3.5 for forwards and 1.75 for defencemen) because GVT is goals versus threshold, not just goals.

2004 DRAFT-ELIGIBLE PLAYERS RANKED BY OFFENSIVE GVT PROJECTION (OGVT)

PLAYER	LEAGUE	EQ PTS	OGVT
Cam Barker (D)	WHL	74	8.95
Andrew Ladd	WHL	82	8.63
Brandon Dubinsky	WHL	81	8.20
Rob Schremp	OHL	82	7.43
Wojtech Wolski	OHL	75	7.36

PLAYER	LEAGUE	EQ PTS	OGVT
Ryan Callahan	OHL	79	7.17
Dave Bolland	OHL	76	6.89
Dylan Hunter	OHL	74	6.54
Peter Tsimikalis	OHL	69	6.35
Rob Hisey	OHL	76	6.24
Chad Klassen	WHL	73	6.17
Alexandre Picard	QMJHL	68	5.90
Drew Stafford	WCHA	61	5.89
Mark Mancari	OHL	63	5.36
Ryan Chipchura	WHL	61	5.07
Josh Gorges (D)	WHL	42	4.87
T.J. Hensick	CCHA	67	4.81
Liam Reddox	OHL	63	4.72
John Vigilante	OHL	63	4.67
Travis Zajac	BCHL	60	4.64

Now that we have an offensive projection, we can move on to the defensive side of the game.

Creating the Defensive Projection

When projecting offensive production, we have numbers to work from in the player's scoring stats. We don't have the same starting point when dealing with defence. This means we need to start with an average baseline and apply adjustments as best we can to arrive at a projection for that particular player.

1. So we start with a baseline defensive GVT of 3.5 for defencemen and 1.0 for forwards.

2. To that we apply the same league-quality multiplier that we did on offence, to reflect the fact that the average defender in a higher-quality league is better than the average defender in a lower-quality league.

3. And since a player's size is presumably more important for defence than for offence, we apply a fairly significant adjustment based on a player's height, moreso for blueliners than for forwards. Again, while we may suspect that many scouts overvalue size, the idea that it is something of an advantage does make sense.

4. To further acknowledge the results of scouting in player evaluation, playing in the World Junior Championship also results in an increase to the defensive projection, and again this adjustment is larger for blueliners, since defence is a much more important aspect of the game for that position.

5. Finally, a threshold amount is deducted to get from goals to goals versus threshold (0.25 for forwards and 2.00 for defencemen).

And there you have it. That's all we have for a defensive GVT projection, abbreviated DGVT. Because there are fewer factors involved, the projections for defence will always be in a much narrower range than the offensive projections. While offence might range from a negative projected GVT up to, say, 20 for the very best prospects, the defensive range might be from zero to seven GVT or so. This reflects actual GVT results as well. In the NHL, the variance in defensive ratings is always narrower than for offence.

Creating the Overall Projection

Now that we have a projection for both offensive GVT and defensive GVT, we can simply add them together to get a projection of the player's overall value. But this process gives only a projection for a single year's data, and for many players we have more than one season's worth of information to develop a projection from. In such a case, it would be intellectually irresponsible to ignore the 2002–03 season, for example, simply because we're projecting for the 2004 draft.

We would expect the more recent season to be more relevant, of course, so to produce an overall rating we use a weighted average of each season's projection for that player. This will use the same approach as that used in the team-building chapter, where each previous season is weighted one-half as much as the current season. So if you have a player who has played three seasons, the current season is given a full weight, the previous season is weighted at one-half, and the season before that is weighted at one-quarter.

As an example, let's take Dennis Wideman in the 2002 entry draft. Wideman was eligible for the draft in 2001 but was not drafted. So in 2002 there were three seasons of data for him, since he played in the OHL in 1999–2000, 2000–01, and 2001–02. Since we have three seasons, they will be weighted at one-quarter, one-half, and one, respectively, for a total weight of one and three-quarters. So his overall projection for 2002 is calculated as follows:

DENNIS WIDEMAN WEIGHTED RATING CALCULATION

SEASON	OGVT	DGVT	RATING	WT	WT OGVT	WT DGVT	WT PROJ
2000	3.21	1.99	5.20	0.25 / 1.75	0.46	0.28	0.74
2001	2.77	1.82	4.59	0.5 / 1.75	0.79	0.52	1.31
2002	5.69	1.69	7.38	1 / 1.75	3.25	0.97	4.22
				TOTAL	4.50	1.77	6.27

But there's another consideration that we have not yet addressed. Wideman is a defenceman, and his projections are intended to represent what sort of GVT he will produce in the NHL as a defenceman. As a blueliner, he will receive more ice time, on average, than a forward, and as such he will have more opportunity to accumulate GVT in excess of the replacement-level threshold. So when we're comparing forwards and defencemen in order to rank them for draft purposes, we need to consider this additional ice time. Otherwise, you're going to end up with a draft ranking dominated by blueliners. As such, any defenceman has his rating scaled down by a factor of 0.8, representing the approximate ratio in ice time between forwards and defencemen.

So in Wideman's case, his time-adjusted OGVT is 3.73 and DGVT is 1.47, resulting in an overall rating of 5.20. This is quite a good rating, deserving of better than the 241st pick, which is where he went in 2002 after being passed over entirely in 2001. The scouts certainly missed on Wideman (he wasn't ranked among the 120 North American skaters evaluated by the CSS for the 2002 draft), and a large part of this was surely due to his relatively small stature and lack of physical play. But the scouts don't miss everything, and ignoring their insights would be a dangerous path to saunter down when evaluating young hockey players.

Synthesizing Statistical Analysis and Traditional Scouting

The Projectinator analysis demonstrates that a good deal of useful information can be extracted from a player's junior hockey stats and that this information can be used to develop fairly accurate projections for a player's NHL career. This should not, in any way, be interpreted to mean that the NHL should do away with their scouting departments and bring nothing but spreadsheets to the next entry draft (despite the obvious cost savings). It's unlikely that any NHL team completely disregards stats in favour of opinion any more, though I suppose some might. The truth is that both approaches have their advantages and disadvantages. Statistical analysis avoids certain biases that human scouts can be subject to, and it can cover pretty much every eligible player without the need to have a scout making several trips to see each individual player. But it does miss some things that scouts can see, obvious flaws that may not hurt in junior but would be exploited in the NHL, and how and where the player can grow his game. Scouts can also spot personality issues, which the numbers cannot.

Scouts and stats very often agree about a particular player's status as a prospect. The best prospects are often rated very highly by both the numbers and the scouts, and on the other end of the spectrum there is a host of eligible players annually who neither rating would consider to be worth drafting. But between these two extremes there is a significant number of young hockeyists on whom there is some degree of disagreement. For these players, synthesizing the results of both types of draft rankings might produce superior results, by controlling for the flaws of one rating with the virtues of the other.

We have already incorporated some amount of scouting acumen into the GVT projections, since we have considered a player's appearances in the World Juniors. But that is *retrospective scouting,* using a player's past performance to identify the best players at the time of the tournament, whereas for our purposes we prefer *prospective scouting,* which attempts to identify which players will be best in the future. For that, we turn to the final annual rankings compiled by the NHL's Central Scouting Service (CSS).*

[* All CSS ranking information obtained from the Draft Analyst, accessed June 19, 2015, http://www.thedraftanalyst.com.]

This gives us two sets of rankings: one resulting from statistical analysis and one from scouting analysis. For a significant proportion of draft-eligible players, these two rankings will be very similar to each other. But there are always exceptions, and sometimes these exceptions represent drastically different opinions between the two rankings. In the 2004 rankings, for example, the statistical analysis suggests that Rob Hisey was the ninth-best North American skater available. Meanwhile, on the scouting side, the CSS ranked 187 North American skaters, and Hisey didn't even place. There could hardly be a bigger disagreement in this context.

There are at least two broad possibilities that result in this disagreement. The first is that the scouts saw something in Hisey's game that they thought would cripple his NHL career, something that statistical analysis is blind to. Maybe his skating wasn't very good or he wasn't defensively responsible, and while he could get away with it in junior, at the professional level the opponents would take advantage of this weakness so much that his value would be negligible. (Or, you know, maybe it's because he's 5-foot-8.) The second possibility is that the scouts thought they saw something like this that wasn't really there, or perhaps they saw something that they thought was very significant but were severely overestimating the importance of (because, you know, he's 5-foot-8). Since we know that scouts often overestimate the value of size, we need to be careful with the degree to which we incorporate the CSS rankings into the final Projectinator rankings.

We know in retrospect that in Hisey's case, the scouts were closer to the mark than the numbers. While Hisey has had a decent pro career, playing some in the AHL but largely in Europe, it's now clear he was not NHL material. But the scouts don't always get it right. They passed on the chance to draft Martin St. Louis in both 1994 and 1995, for example, despite his having numbers that screamed "draft me." So while we should consider scouting opinion, hopefully to avoid rating players such as Hisey too highly, we shouldn't simply defer to it.

THE ADJUSTMENT ITSELF IS SIMPLE

First we compare a player's CSS ranking to his ranking in the GVT projections. Since the CSS ranked 187 North American skaters in 2004, we consider any eligible player not on their list to be ranked 188th. Similarly, any player beyond the top 187 in GVT projections is ranked 188th for this purpose.

1. We then take the difference between the two rankings and apply an exponent of 0.7, in order to ensure that we are considering the CSS rankings but not deferring to them. This prevents the scouting adjustment from overwhelming the statistical analysis for players on whom there is significant disagreement; if we allow the scouting results to dominate here, we might as well not be doing the analysis at all.
2. Then for each adjusted point of difference in the rankings, we add or subtract a bit of projected GVT in the appropriate direction. Forwards gain or lose 0.04 GVT and 0.035 DGVT for each point of difference, while defencemen gain or lose 0.02 OGVT and 0.05 DGVT for each point of difference. This is intended to reflect the relative importance of offence and defence to each position and also to recognize the fact that if scouts significantly disagree with the numbers, it's more likely that it's a defensive issue than an offensive one, since the defensive side is not nearly as well-represented in the numbers as the offensive side.

FINAL RESULTS

So there we have it. With this final step we have developed a statistical rating system that also incorporates scouting opinion in an attempt to take the best of both worlds and avoid the worst of both as well. The following table presents the top 50 North American skaters in the 2004 entry draft ranked by their GVT projected by the Projectinator (Proj). Each player's CSS ranking (among players covered by the Projectinator) and his actual draft position (again among this set of players) are also provided for comparison.

TOP 50 DRAFT-ELIGIBLE NORTH AMERICAN SKATERS BY PROJECTINATOR RATING, 2004*

RANK	PLAYER	POS	LEAGUE	CSS RK	DRAFT RK	OGVT	DGVT	PROJ
1	Cam Barker	D	WHL	2	1	6.66	1.49	8.15
2	Rob Schremp	F	OHL	10	11	7.19	0.63	7.82
3	Wojtech Wolski	F	OHL	5	9	6.85	0.86	7.71
4	Drew Stafford	F	WCHA	7	6	5.80	1.66	7.46
5	Andrew Ladd	F	WHL	1	2	6.09	0.89	6.98
6	Brandon Dubinsky	F	WHL	76	33	5.28	0.17	5.45
7	Dave Bolland	F	OHL	8	16	4.31	0.97	5.28
8	Peter Tsimikalis	F	OHL	33	–	4.54	0.70	5.24
9	Kyle Chipchura	F	WHL	4	7	3.87	1.22	5.09
10	Alexandre Picard	F	QMJHL	3	3	4.07	1.01	5.08
11	Mike Green	D	WHL	9	14	3.03	1.95	4.98
12	Jeff Schultz	D	WHL	12	12	1.98	2.96	4.94
13	Mike Lundin	D	HE	53	51	1.56	3.29	4.85
14	A.J. Thelen	D	CCHA	11	5	2.36	2.47	4.83
15	Wes O'Neill	D	CCHA	22	59	1.27	3.42	4.69
16	Ryan Callahan	F	OHL	72	66	4.33	0.18	4.51
17	Adam Berti	F	OHL	21	38	3.30	1.15	4.45
18	Travis Zajac	F	BCHL	15	8	3.39	0.89	4.28
19	Kyle Wharton	D	OHL	13	32	1.09	3.08	4.17
20	Liam Reddox	F	OHL	90	57	3.87	0.25	4.12
21	Bruce Graham	F	QMJHL	14	28	2.77	1.31	4.08
22	Boris Valabik	D	OHL	6	4	0.85	3.23	4.08
23	Ryan Garlock	F	OHL	16	24	2.60	1.47	4.07
24	Evan McGrath	F	OHL	35	67	3.25	0.76	4.01
25	Rob Hisey	F	OHL	–	–	4.10	–0.12	3.98
26	Grant Lewis	D	ECAC	20	20	1.50	2.48	3.98

RANK	PLAYER	POS	LEAGUE	CSS RK	DRAFT RK	OGVT	DGVT	PROJ
27	Bryan Bickell	F	OHL	17	21	2.20	1.72	3.92
28	David Booth	F	CCHA	26	30	2.38	1.52	3.90
29	Vaclav Meidl	F	OHL	27	45	2.68	1.22	3.90
30	David Booth	F	CCHA	26	30	2.38	1.52	3.90
31	Grant Lewis	D	ECAC	20	20	1.45	2.41	3.85
32	Kris Chucko	F	BCHL	28	10	2.87	0.94	3.82
33	Mike Funk	D	WHL	38	22	1.56	2.17	3.73
34	Aki Seitsonen	F	WHL	29	60	2.31	1.38	3.69
35	Roman Tesliuk	D	WHL	25	23	0.77	2.82	3.60
36	Chad Painchaud	F	OHL	41	52	2.47	1.09	3.56
37	Martins Karsums	F	QMJHL	40	35	2.73	0.81	3.54
38	Chris Zarb	D	USHL	36	75	1.25	2.27	3.52
39	Casey Corer	D	WCHA	51	39	1.00	2.49	3.49
40	Michal Sersen	D	QMJHL	42	68	1.73	1.74	3.47
41	Victor Oreskovich	F	USHL	70	31	1.84	1.61	3.45
42	Jake Dowell	F	WCHA	61	72	1.38	2.05	3.43
43	Tyler Haskins	F	OHL	45	84	2.45	0.99	3.43
44	Jordan Smith	D	OHL	50	19	1.09	2.33	3.42
45	Clayton Barthel	D	WHL	52	47	0.69	2.72	3.41
46	Brian Inhacak	F	ECAC	43	138	1.91	1.50	3.41
47	Kyle Wilson	F	ECAC	73	143	1.76	1.64	3.40
48	Steve Birnstill	D	HE	60	–	0.69	2.70	3.39
49	Bret Nasby	D	OHL	74	79	0.34	3.05	3.39
50	John Lammers	F	WHL	47	46	2.22	1.14	3.37

[* All CSS ranking information obtained from the Draft Analyst, accessed June 19, 2015, http://www.thedraftanalyst.com.]

It's plain to see that for a big chunk of these players, the differences between the CSS rankings and their Projectinator rankings are negligible. In comparing the CSS and Projectinator rankings for these 50 players, there are maybe 34 for whom any difference is insignificant, 11 for whom any difference is moderate, and five for whom the difference is significant, depending on how you define what is significant. And there are some players with a difference in the rankings that the Projectinator does a better job with (based on their future career) and some that the scouts do a better job with.

Overall, looking at all players and not just the top-ranked ones, the Projectinator seems to project junior hockey prospects somewhat better than the CSS rankings do (bearing in mind that the CSS rankings are incorporated into the Projectinator rankings, so they're not completely independent). The great majority of players are actually ranked very similarly by both methods, but if you take the players that the Projectinator projects significantly better than the CSS rankings and compare them with the players it projects significantly worse, there are probably about 5–10% more "betters" than "worses." Moreover, the hits include several very good NHL players, while the misses don't include many significant players. That is, there are a number of good NHLers (such as Brandon Dubinsky, Ryan Callahan, and Dan Girardi as well as lesser players like André Benoit and Chris Campoli) that the Projectinator ranks much higher than CSS does, while there's really only one good NHLer (Mason Raymond) that the CSS rankings do a much better job with. Of course, this may be a fluke since we're looking at only a single season and because the model was built using these players, but it is interesting nonetheless. It's also interesting that the players the Projectinator does better on are mostly defencemen, plus, in Dubinsky, a forward who has a good defensive reputation. One would think the scouts would do better on defence while the numbers would do better on offence, but there you go.

We should also remember that the Projectinator uses only the most basic stats in its model: games played, goals, assists, points. An NHL team would, of course, have the resources to acquire more detailed statistics. Even if you had only power-play scoring numbers for every league, the quality of your offensive GVT modelling should improve. For example, a big part of Rob Schremp's draft year success was that 63% of his scoring came with the man advantage, but we don't have that data readily available for the wide range of leagues under study, so we do the best we can with what we have.

You might notice how tight the projected GVT values are and how low the highest values are compared to the actual best GVTs in the NHL each year. This is not a fluke and not, by itself, an indictment of the 2004 North American draft class. It's actually the result of the methods we use to build the projection model. Whenever you build a statistical model by aggregating data in the way we have, the specific results of the model will tend to be conservative (that is, tending toward the middle of the scale). Ultimately the best players will tend to exceed their projections to a significant degree, while the worst will fall far short of them, often recording a negative career GVT.

Performance of individual players in the NHL tends to follow a log-normal distribution. That is, there are a small number of players who perform very badly (small because players who do so do not last in the league) and then a much larger number who perform in the middle range, which gradually decreases as the level of performance increases, ending with only a very small number of players at the very highest level, a level that is far more above average than the worst performers are below average. This distribution is illustrated below with the black line, while the grey line shows what the projections for this group of players would look like. The projection is also a log-normal distribution, but its values have a much tighter range, and far more players are grouped in the middle of the performance range. This is simply the nature of models such as the Projectinator.

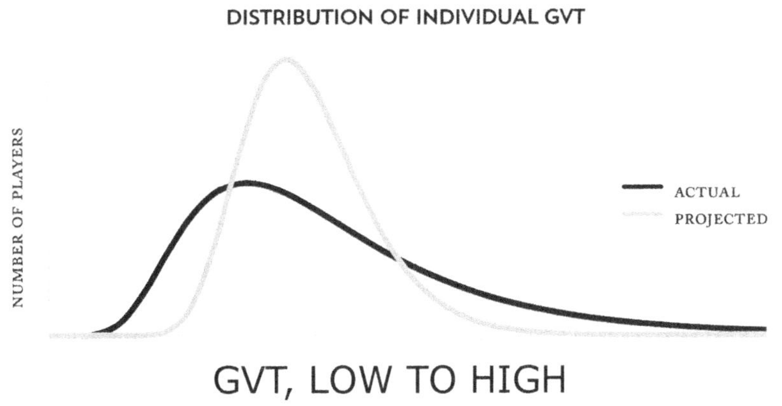

We should put the 2004 results into some kind of context. To do so, we can look at the top-rated players from a variety of seasons. Let's start with 2002, since that's the earliest of the consecutive years for which the relevant CSS rankings are widely available at the time of writing. How does the top of the 2004 North American draft class compare to other years?

PROJECTINATOR RESULTS FOR NUMBER-ONE CSS-RANKED NORTH AMERICAN SKATERS

YEAR	PLAYER	POS	OGVT	DGVT	PROJ
2002	Jay Bouwmeester	D	5.96	2.68	8.64
2003	Eric Staal	C	8.46	1.04	9.50
2004	Andrew Ladd	C	6.09	0.89	6.98
2005	Sidney Crosby	C	22.85	0.95	23.77
2006	Erik Johnson	D	1.65	2.59	4.24
2007	Kyle Turris	C	8.18	0.67	8.85
2008	Steven Stamkos	C	13.21	1.14	14.35
2009	John Tavares	C	20.18	1.17	21.35
2010	Tyler Seguin	C	10.32	0.92	11.24
2011	Ryan Nugent-Hopkins	C	9.07	0.83	9.90
2012	Nail Yakupov	C	9.01	1.10	10.11
2013	Seth Jones	D	5.12	3.17	8.29
2014	Sam Bennett	C	7.64	1.02	8.66
2015	Connor McDavid	C	20.09	1.20	21.29

Of these 14 drafts, the Projectinator disagrees with the CSS ranking exactly half the time. The best North American skater prospect in 2002 according to the Projectinator was Rick Nash, with a rating of 12.08 projected GVT. No disrespect to Bouwmeester, but Nash has certainly had the superior NHL career, and indeed in 2002 the Columbus Blue Jackets agreed with this assessment and selected Nash first overall. In 2004, Cam Barker was the best prospect, ahead of Andrew Ladd. It seems clear at this point that Ladd has been the better player, so the CSS wins that one.

In 2006 the CSS thought Erik Johnson was the best prospect, and the St. Louis Blues agreed. The statistical analysis strongly disagrees with that assessment, ranking Johnson around the bottom of the first round, not the top. The Projectinator instead recommends drafting Jonathan Toews in 2006 (rating of 9.13 GVT), which Chicago did with the third overall pick, and I'm sure they're pretty happy with that selection. In 2007 the CSS recommended Kyle Turris as the best available North American skater. Turris has recently made progress toward a solid NHL career, but Chicago disagreed with the CSS in this case, instead taking Patrick Kane, who is the player that the Projectinator would recommend (rating of 13.03 GVT). Again, I suspect the Hawks are content with their choice.

In 2010, Taylor Hall is rated significantly ahead of Tyler Seguin (18.01 GVT to 11.24 GVT). Hall was ranked by the CSS ahead of Seguin for most of the season leading up to the 2010 draft, but their final 2010 rankings moved Seguin ahead. While it's still too early to tell, a betting man would probably put his money on Seguin at this point, though the difference probably won't be huge in the end.

The top-rated player of 2013 by this system is Nathan MacKinnon (11.61 GVT), and in 2014 it's Sam Reinhart (13.43 GVT), and both of these drafts are too recent to say anything about the results yet, although it's worth noting that Florida, Buffalo, and Edmonton all disagreed with the CSS ranking in 2014, taking Aaron Ekblad, Reinhart, and Leon Draisaitl ahead of Bennett, who went fourth to Calgary.

It's kind of a shame that we picked 2004 as the focus year for this study, because it appears to be the weakest year for North American skaters in recent memory. The 2004 entry draft was topped by Alex Ovechkin and Evgeni Malkin, but the North American prospects were a bit weak. It's the only draft since 2002 in which the top North American skater has a rating of less than 9.00 GVT. The usual top rating is somewhere between 9.50 and 14.50 GVT, with ratings in excess of that being rare.

Another way to look at the 2004 results is to examine the best NHL players who were available at the 2004 entry draft and compare their projected and actual GVTs. These are the players whom you ultimately want to get right, as best you can.

HIGHEST GVT AMONG NORTH AMERICAN SKATERS AVAILABLE IN 2004 ENTRY DRAFT

PLAYER	DRAFT RK (Y)	PROJ	AVG GVT TO AGE 27	GVT, BEST 2 YEARS	GVT RATING
Mike Green	29 (2004)	4.98	9.42	24.80	17.11
Paul Stastny	44 (2005)	3.57	10.36	16.20	13.28
Travis Zajac	20 (2004)	4.28	6.81	14.30	10.56
Andrew Ladd	4 (2004)	6.98	8.04	12.75	10.40

PLAYER	DRAFT RK (Y)	PROJ	AVG GVT TO AGE 27	GVT, BEST 2 YEARS	GVT RATING
Alex Goligoski	61 (2004)	1.70	7.23	11.90	9.56
Kris Versteeg	134 (2004)	0.65	6.00	11.50	8.75
Drew Stafford	13 (2004)	7.47	6.34	11.10	8.72
Wojtek Wolski	21 (2004)	7.72	6.35	11.05	8.70
Dan Girardi	undrafted	2.54	7.15	10.15	8.65
Ryan Callahan	127 (2004)	4.52	6.30	10.80	8.55
David Desharnais	undrafted	1.20	5.58	10.35	7.97
Jason Garrison	undrafted	1.46	6.87	9.05	7.96
Brandon Dubinsky	60 (2004)	5.45	6.00	9.80	7.90
David Booth	53 (2004)	3.90	3.97	10.40	7.18
Mason Raymond	51 (2005)	-0.70	5.11	9.05	7.08
Dave Bolland	32 (2004)	5.28	4.04	9.15	6.59
Tyler Kennedy	99 (2004)	3.31	5.22	7.90	6.56
Jeff Schultz	27 (2004)	4.94	3.49	8.40	5.94
Chris Campoli	227 (2004)	2.45	4.57	6.70	5.64
Blake Comeau	47 (2004)	3.15	2.64	7.80	5.22
Bryan Bickell	41 (2004)	3.92	2.94	7.05	5.00
Cam Barker	3 (2004)	8.15	2.44	7.25	4.85
Troy Brouwer	214 (2004)	3.03	2.75	6.80	4.78
Mike Santorelli	178 (2004)	1.33	1.78	5.30	3.54
Jack Hillen	undrafted	0.69	2.13	4.95	3.54
Rob Schremp	25 (2004)	7.82	2.35	4.70	3.53
Daniel Winnik	265 (2004)	2.18	2.12	3.90	3.01
Mike Lundin	102 (2004)	4.86	1.98	3.90	2.94
Brandon Prust	70 (2004)	0.73	1.60	4.15	2.88
Mark Fistric	28 (2004)	3.10	1.53	4.15	2.84
Dustin Boyd	98 (2004)	3.34	1.22	3.30	2.26
Brandon Yip	239 (2004)	2.45	1.90	2.60	2.25
Michael Kostka	undrafted	1.13	2.10	2.10	2.10
Jake Dowell	140 (2004)	3.43	1.57	2.05	1.81
Kyle Chipchura	18 (2004)	5.09	0.76	2.45	1.60
Davis Drewiske	undrafted	1.30	1.30	1.50	1.40
Nate Prosser	undrafted	0.66	1.08	1.65	1.36
Alexandre Picard	8 (2004)	5.09	0.33	2.25	1.29
Lee Sweatt	undrafted	1.02	1.20	1.20	1.20

PLAYER	DRAFT RK (Y)	PROJ	AVG GVT TO AGE 27	GVT, BEST 2 YEARS	GVT RATING
Jordan Hendry	undrafted	0.80	0.65	1.65	1.15
Patrick Kaleta	176 (2004)	1.54	0.23	1.40	0.81
T.J. Hensick	88 (2005)	3.40	0.45	1.10	0.78
Bryan Rodney	undrafted	2.81	0.50	0.85	0.68
Brett Festerling	undrafted	2.58	0.20	0.70	0.45
Aaron Gagnon	240 (2004)	1.22	0.13	0.60	0.36
Mark Mancari	207 (2004)	1.74	0.12	0.55	0.34
Danny Syvret	81 (2005)	1.77	-0.08	0.70	0.31
Mike Moore	undrafted	0.37	0.20	0.20	0.20
Tim Wallace	undrafted	1.02	0.20	0.20	0.20
Robbie Earl	187 (2004)	3.20	0.03	0.35	0.19

If you're comfortable digesting numbers in table format, you can see that there is a relationship between the projected and actual GVT here, suggesting that the Projectinator model does a pretty good job of forecasting the future. But if you're more visually oriented, the following point chart summarizes the data, plotting each player's actual GVT against its projection.

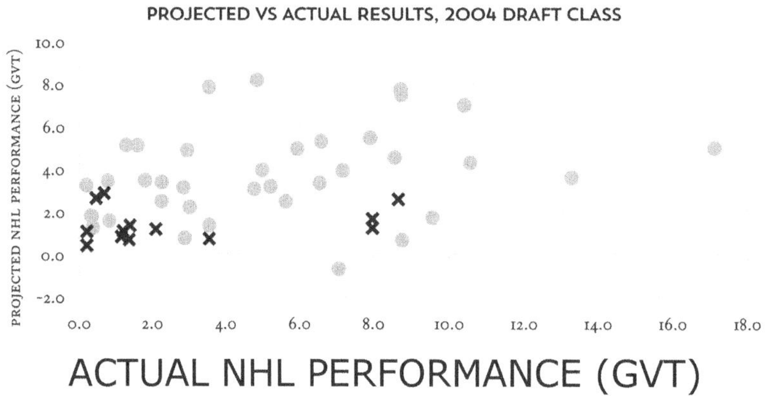

You can see there is a slight upward slope to the data, reading from left to right, which again is an indication of a positive relationship between the two sets of data: the higher the player's projection, the higher his actual GVT. It's also worth noting the players marked with an "X" in the chart; these are the undrafted players, the ones that no team thought worthy of being drafted. Indeed, if you look at the draft information in the table, you can see that the relationship between a player's draft placing and his actual GVT is positive but not strongly so. So again, the Projectinator is far from perfect, but the same can be said of opinion-based drafting.

Lottery Winners: Edmonton Gets McDavid

The 2015 NHL draft lottery produced a dramatic win for the Edmonton Oilers, who moved from the third overall pick to first, thereby winning the Connor McDavid sweepstakes. Picking third, the CSS projected that the Oilers would probably have ended up with Boston College defenceman Noah Hanifin. Meanwhile, the Projectinator would say it was Jack Eichel, the Boston University centre who was ranked second by the CSS. Either of those players would have been just fine and dandy to a team picking third, but by moving up to the number-one spot, the Oilers were able to select a franchise player in McDavid.

No one had any question about McDavid's talent going into the 2015 draft. He drips talent and hockey acumen. The question was not whether he was one of the greatest prospects of the draft era, but where he ranked among the all-time greats. As Jonathan Willis pointed out in a pre-draft article, even Wayne Gretzky was on the McDavid bandwagon, calling him the best prospect in the last 30 years, a clear reference to Mario Lemieux, who was drafted in 1984.* Some said that McDavid would be a better player than Crosby, though ultimately Willis disagreed, using a simple analysis to peg McDavid's junior numbers as being superior to Eric Lindros's and John Tavares's but below those of Crosby and Lemieux.

[* Jonathan Willis, "How Connor McDavid Stacks Up Against the All-Time Greats," *Bleacher Report,* June 23, 2015, http://bleacherreport.com/articles/2504597-how-connor-mcdavid-stacks-up-against-thealltime-greats.]

The Projectinator system is not intended to directly compare players from different seasons; it is specifically meant to rank players from the same season. And since the model was built using data from the 2000s, the further back you go in time, the more likely it is that circumstances were different enough that direct comparisons begin to break down. As such, comparing players from different times using the Projectinator's results is more fraught with uncertainty than usual. Just keep that in mind as we look at the top entry draft prospects of all time, according to the Projectinator.

TOP FIVE ALL-TIME PROJECTINATOR RATINGS FOR NORTH AMERICAN SKATERS

RANK	PLAYER	DRAFT Y	OFF	DEF	PROJ
1	Sidney Crosby	2005	22.82	0.95	23.77
2	John Tavares	2009	20.18	1.17	21.35
3	Connor McDavid	2015	20.09	1.20	21.29
4	Mario Lemieux	1984	19.59	0.90	20.49
5	Eric Lindros	1991	18.29	1.34	19.63

The comments about comparisons across time are applicable to Lemieux and Lindros. They are so long ago that direct comparisons using this system are probably not valid. If a 1980s version of the Projectinator were developed and used for recent players as well, Lemieux would quite possibly be at the top of the list here. On the other hand, it may be worth noting that Wayne Gretzky's 1977–78 season (which predates the entry draft) actually has a rating of 25.67 GVT, and that was one season before he would be draft-eligible today. The best players generally have a better rating in their draft year than the year before, so we would expect Gretzky to have had a rating of maybe 28 to 30 had he stayed in junior rather than circumventing the draft by signing with the WHA. And the circumstances that he played in, when even the best players remained in junior hockey until the age of 20 or so, means he was probably facing tougher competition. So like Lemieux and Lindros, his rating would probably be even higher if the projections were calibrated for players from his own time period. This does offer some insight into the question of whether there has been a player to come along since Wayne Gretzky who was as good a prospect as he was. The answer to that is pretty clearly no, and it's not close.

The point about players generally getting better may be relevant to John Tavares's ranking. While it's hard to say he hasn't produced at the NHL level, he hasn't really met the expectations that his junior numbers suggested. Two seasons before his draft year, his Projectinator rating was 25.35. The season before his draft year, it was 24.10. But in his draft year it was down to 19.70. Since the final Projectinator rating is based on a weighted average of these seasons, Tavares ends up with a rating over 21, but the fact that his rating declined each season may indicate that his earlier seasons are not relevant compared to the last one. You generally expect top prospects to be improving each year, and a player who instead declines might suggest someone with less upside than you would otherwise suspect. Sidney Crosby's ratings were 21.60 in 2003–04 and 24.85 in 2004–05, which is a fairly normal progression. So while the Projectinator does not, at present, consider changes such as this when calculating its rankings, it may be sufficient reason to conclude that Tavares was not quite as good a prospect as the table might make him seem. But even if his projected GVT was only 19.70, that's still eye-popping.

McDavid certainly does not show this decline. His ratings in the three seasons leading up to the draft are 11.29, 18.70, and 25.09, so he progressed by leaps each year in junior. As such, there are reasons to maybe rank him ahead of Tavares, but what about Crosby? Willis concluded that Crosby's junior numbers were more impressive than McDavid's,* and it's easy to agree with that assessment quite strongly. So maybe Crosby was a better prospect than McDavid is. But while some people think of Crosby as something of a disappointment given the hype he endured during his junior years, this is a player who before his 27th birthday had won two NHL scoring titles and two Hart Trophies as league MVP, made four All-Star Teams at centre, and captained a Stanley Cup–championship team. So if the verdict on McDavid is "not quite as good a prospect as Sidney Crosby," then you'll forgive the Oilers for their chortling. There's every reason to believe he will be the best player in the NHL in a few years.

[* Jonathan Willis, "How Connor McDavid Stacks Up Against the All-Time Greats," *Bleacher Report,* June 23, 2015, http://bleacherreport.com/articles/2504597-how-connor-mcdavid-stacks-up-against-thealltime-greats.]

The Sabres probably don't feel quite as good about the results of the draft lottery, of course. But Jack Eichel is quite a good consolation prize for Buffalo. According to his Projectinator ranking, in some past seasons he would have easily been the best player available. And who knows, maybe Buffalo would have drafted Eichel even if they had had the first overall pick in 2015. Indeed, there were some people who suggested that maybe Eichel should have been rated ahead of McDavid to begin with, arguing that he had some significant advantages over the Erie Otter.

Philip Wischmeyer, for example, wrote that "McDavid has the superior speed, quickness, stick handling and vision, but Eichel has been thought to have slightly better size, determination, competitiveness and leadership abilities, while also being extremely talented in his own right."* You might notice that all of the advantages for McDavid are physical and largely measurable, while the advantages attributed to Eichel are, for the most part, what are known as "intangibles," so perhaps an extra-healthy dose of skepticism is due here. Many people do have a tendency to retreat to intangibles when they *just feel* that one player is better than another. This is why, while we do consider the CSS rankings in the Projectinator (because there are surely intangibles that matter to some degree), we try not to put too heavy a weight on them (because scouts often have trouble evaluating the impact that such intangibles actually have on the ice).

[* Philip Wischmeyer, "Could Jack Eichel Dethrone McDavid as the Top Pick?," *Along the Boards* (blog), June 19, 2015, http://alongtheboards.com/2015/06/could-jack-eichel-dethrone-mcdavid-as-the-top-pick/.]

As it turns out, the system even ranks Dylan Strome ahead of Eichel in the 2015 draft (12.34 projected GVT to 10.34). One thing this version of the

Projectinator does not consider is a player's teammates. Strome surely benefited from having Connor McDavid on his team. While Strome and McDavid did not play together at even strength, they did on the power play, and the fact that the other team's best checkers would generally be assigned to McDavid freed up the ice a bit for Strome when he was centring the Erie second line. It's also easy to overestimate the effect that McDavid had, as writers such as Mark Scheig have pointed out.* While Strome is seen as a legitimate future NHL star by many people, most agree Jack Eichel should be better. Ultimately, only time will tell. Due to the uncertainty inherent in projections such as these, the difference between a projection of 10 and a projection of 12 is very small.

[* Mark Scheig, "Dylan Strome: Great Player or Trendy Draft Pick?," *The Hockey Writers,* April 9, 2015, http://thehockeywriters.com/dylan-strome-great-player-or-trendy-pick/.]

Future Work

The Projectinator is a solid system built on very simple hockey stats. It's nowhere near perfect, and it could never be. But even in the context of its intent, there is always room for improvement. A couple of factors that might make for improvement were discussed already: the quality of a player's teammates, and whether the player declines in performance year to year in junior.

There are other possible improvements. A model like the Projectinator should always be a work in progress, building in adjustments as new information becomes available or as new analyses are performed. Improvement would likely result if we were able to incorporate situational scoring stats into the analysis; if we had power-play scoring information for all leagues, that would probably provide more accurate results. Since man-advantage scoring is situational, it relies more on coaching preferences, since every player gets some even-strength time while only a small number of players get the bulk of the power-play time.

For defencemen, there may be reason to believe that players with a wide disparity between the offensive and defensive projections may be less likely to pan out than players with a more balanced projection. Cam Barker was very offence-heavy, and he did not have the career that his draft ranking would have indicated. Meanwhile, Josh Gorges was more balanced between the two and exceeded his draft expectations. So that would be something to consider, though more study would be needed to reach a conclusion.

The defensive projections as a whole might be a bit low as well, especially among forwards. There is little variation in forward defensive projections, so that could be improved as well in an attempt to identify the best defensive forwards available. Defensive projections could probably also be improved by lessening the effect of league quality, since NCAA players are probably rated too highly there. Also, giving defensive credit for playing in the World Juniors should probably factor in a player's age; it's more impressive for a 17-year-old blueliner to make a national team than an 18-year-old. Age adjustment for the defensive GVT projection might be valid on the whole. There's nothing to be ashamed of in needing to make improvements to your model. Indeed, in analysis like this a refusal to make adjustments to your work based on further analysis is a sign that you have too much confidence in your results. You always need to be ready to admit that your work may have holes in it that need patching.

Some readers may be familiar with the work I did with an earlier version of the Projectinator online, several years ago.* In the earlier, less-rigorous version of the system, three other factors were included that should perhaps be included here as well. Penalty minutes had a negative effect on a player's projection, as did having only a single year's worth of data to work with for that player (in recognition of the increased uncertainty). Also, it seemed that European players who came over to play North American junior hockey did need a bit of time to adjust to the different style of play. All of these things may be incorporated into this new system in the future. As I said, the work is never really done.

[* Iain Fyffe, "Up and Coming: Predicting NHL Success," *Hockey Prospectus* (blog), June 9, 2009, http://www.hockeyprospectus.com/puck/article.php?articleid=172.]

All of these potential improvements are for the model for the group of players we've been considering in this chapter. Despite this already being quite an extensive discussion, there are many more players who need to be considered as well. Here we have taken a deep look into a narrow subset of players, but to really build a solid model you need to expand your view, applying the same depth to a wide range of players. European skaters are obviously extremely important, and while the approach to projecting their careers would be built on this one, there is one wrinkle that we didn't have to consider with the North Americans. In Europe, many draft-age players play in junior leagues, but the best of them often play in senior leagues in their draft years, 17-year-olds playing against fully matured hockey players. This introduces a new challenge to the projections, but we're up for it. And then we have to consider the goaltenders as well, both North American and European. Goaltenders are much more of a crapshoot when compared to skaters, since their junior numbers are more heavily influenced by team performance, but there are methods to consider that.

Finally, when we have all of this work under way, a more thorough analysis of the system's accuracy is called for, covering multiple seasons, to ensure that the model is the best that we can make it. By studying both good seasons and more unpredictable seasons, you improve the overall accuracy, which is what you need if you want to integrate your draft projections into a team-building model (like Rob's). We've presented the foundations of a player projection system here, but there's plenty of work left to do.

Closing Thoughts

Michael Lewis's 2003 book *Moneyball* told the story of Oakland Athletics GM Billy Beane bringing more statistical analysis into baseball management.[9] A good deal of the book was dedicated to the 2002 June amateur draft, when Oakland held six first-round draft picks due to a variety of free-agent losses they sustained, and they had decided to largely overrule their scouts and use statistical data to select players. They were able to draft a large proportion of the players they wanted to get based on their analysis, and Lewis portrayed this as a great victory for Beane and his assistant Paul DePodesta, with the team drafting Nick Swisher, Joe Blanton, John McCurdy, Ben Fritz, Jeremy Brown, Steve Obenchain, and Mark Teahen in the first round.

Lewis played the part of a cheerleader rather than a dispassionate observer with respect to the draft

---

9 Michael Lewis, Moneyball: The Art of Winning an Unfair Game (New York: W.W. Norton & Company, 2003).

results. A short epilogue described Jeremy Brown's early success in minor-league ball, but since the book was published only a year after the draft, there was no time to follow up on these players as they progressed through their careers. If the book had been published later, it could have discussed how, essentially, every player the A's were interested in, and that other teams were not interested in, failed to pan out. John McCurdy played four seasons in the minors, never higher than AA. Ben Fritz played until 2010 and had a part-season in AAA, but he retired with a career 4.76 ERA in the minor leagues. Jeremy Brown had a cup of coffee with the A's and had some decent minor-league numbers, but he was done in 2007. Steve Obenchain was a pitcher who never reached AAA and retired the same season that Brown did. Mark Teahen was the only one who had something of a major-league career, playing 831 games over seven seasons. The only real successes the A's had in the 2002 draft were players that other teams, using traditional scouting methods, were also interested in: Nick Swisher and Joe Blanton. So in the end, Oakland's experience at the 2002 draft was evidence against the idea that a baseball draft could be effectively done by statistical analysis alone, not for it.

That's baseball, of course, not hockey. And while baseball generally has better statistical data available, especially for young amateurs, it also has to cover a great deal more players in many more leagues (there are *how many* NCAA division I baseball conferences?), so it's not clearly in a superior or inferior position to the same approach in hockey. As such, the lessons from the 2002 A's draft probably have at least some applicability to the NHL, most notably, don't dismiss your scouts just yet. As much as we can now do with statistical analysis in hockey (and that's quite a lot), there are still huge gaps in our knowledge. So the best we can do when projecting the careers of young hockey players is to combine analytics with scouting, with the strengths of one compensating for the weaknesses of the other in order to produce the best results possible. That was likely Billy Beane's mistake: he allowed analytics to override scouting rather than to act as a complement to it.

Projecting the careers of junior hockey players is a vital part of team building. As you've seen, young players are far less expensive than veterans due to the collective bargaining agreement, and as such if you can build a team with young players you will be working with a good deal more economic efficiency. Of course, the big advantage that veterans provide is that they're more of a known commodity—you can see how they have performed in the NHL over time, and you have a much better idea of what you're getting when they sign on the dotted line. This is why analysis of junior hockey players is so important; it's an avenue to improve the projection of young hockey players beyond what is currently done. And the Projectinator is not the only system to attempt such a thing. The Projection Project and the Prospect Cohort Success model also attempt to predict a player's future career.* Both of those systems are built on finding comparable players to the player in question and using the results of those other players to predict the future, so they take a different approach to the Projectinator, and it would probably be wise to examine various different analyses rather than just one when making decisions. But one way or another, statistical analysis deserves a place at the table.

[* Lochlin Broatch and Arthur Wheeler, the Projection Project, accessed June 19, 2015, http://www.theprojectionproject.com/; Josh Weissbock, "Draft Analytics: Unveiling the Prospect Cohort Success Model," Canucks Army (blog), May 26, 2015, http://canucksarmy.com/2015/5/26/draft-analytics-unveilingthe-prospect-cohort-success-model.]

WHO IS THE BEST FACEOFF SPECIALIST?

by ROB VOLLMAN

At first glance, the quest for the league's best faceoff specialist may not appear to be as ambitious a topic as when we hunted for the best goalie, goal scorer, playmaker, or penalty-killer, but it requires a perspective that's a good example of the future of hockey analytics.

In essence, a hockey game can be described as a series of puck battles, arguably thousands of them. They occur in various forms all over the ice and in all manpower situations, and involve any combination of the 10 skaters and two goalies, each of whom have conflicting and often unclear objectives. Through their confusing but significant relationships with the battles immediately before and after, these countless battles determine the game's end result, despite the fact that they're resolved almost as frequently by varying degrees of random variation or, possibly, one of the four on-ice officials as they are by the players themselves.

The chaotic nature of these battles is what often confounds our attempts to study the game analytically, especially compared to the simple pitcher-vs-batter duel in baseball. However, there is one common type of puck battle that is always recorded, occurs one on one, without advantage to one side or the other, and has a clear relationship with scoring: the faceoff. Not only can the insights gained from studying this one type of puck battle shed some light on other types of battles, but the same basic objective approach and analytic process can be applied elsewhere.

Given that only 3.3% of scoring chances directly result from this one type of puck battle, according to an estimate by the *Edmonton Journal*'s David Staples, there are still plenty of battles left to cover.* Already there have been statistical hockey analysts who have conducted high-profile research and earned front-office positions by tracking certain battles manually, such as defensive zone breakouts, offensive zone entries, and passing. Once new player-tracking technology makes more data available, this could be the next area of research to really explode.

[* David Staples, "Does the Hockey World Place Too Much Emphasis on Faceoff and Zonestart Percentages?," *Edmonton Journal,* January 24, 2015, http://blogs.edmontonjournal.com/2015/01/24/boydgordon-is-a-faceoff-ace-for-what-its-worth/.]

Another attribute that makes studying faceoffs such an appealing topic is the popular belief that traditional views and measurements are already quite accurate. On the surface, faceoff-winning percentage appears to provide all the answers, without the need for additional statistical analysis. However, a closer examination will reveal that there is a far better way of measuring success in the faceoff circle than a scorekeeper's subjective opinion of who touched the puck first.

Regardless of how faceoff success is measured, there are a lot of outside factors that can influence the results. A player's linemates, the abilities and handedness of his opponent, being on home ice, and having the man advantage are all notable variables that can affect the outcome of a faceoff to one extent or another.

Even once the best faceoff specialists are identified, the question of how much faceoff success truly matters still needs to be addressed. What is the actual relationship between a faceoff win and possession, not to mention the bottom line of goals and wins? And given the various factors and random variation involved, just how large an advantage do the league leaders possess over replacement-level options? It's not enough to simply identify the league's leaders, because if the results aren't placed in the right context, they can never be more than interesting barroom trivia, at best.

In our hunt for the league's best faceoff specialist, we'll start with the traditional view and assign an approximate value to the league's top options relative to replacement level. After exploring the shortcomings of this approach, we'll unveil the new measurement used within the hockey analytics community. Finally, after an exploration of the seven most significant factors that can influence the results, we'll crown our champion. Spoiler alert: it won't be a shocker.

The Traditional View

In the world of statistics, the obvious and traditional approach usually has pretty decent accuracy, especially in areas as relatively baseball-like (for lack of a better term) as faceoffs. Of course, this book isn't being written for those who want only "decent" accuracy.

In this case, a list of the players with the best faceoff percentage (FO%) over the past three seasons combined produces familiar names, like Boston's Patrice Bergeron, Nashville's giant Paul Gaustad, and Chicago's Jonathan Toews, mixed in with a number of defensive-role players who are known to have a similar edge on the draw, like Montreal's Manny Malhotra and Adam Hall, who is currently playing in Switzerland.

TOP 50 FACEOFF SPECIALISTS, 2012–13 TO 2014–15*

PLAYER	TEAM	WON	LOST	FO%
Zenon Konopka	MIN/BUF	507	327	60.8%
Patrice Bergeron	BOS	2,739	1,828	60.0%
Vladimir Sobotka	STL	789	530	59.8%
Manny Malhotra	VAN/CAR/MTL	1,167	789	59.7%
Rich Peverley	BOS/DAL	559	381	59.5%
Adam Hall	TBL/CAR/PHI	455	323	58.5%
Jim Slater	WPG	756	543	58.2%
Paul Gaustad	NSH	1,583	1,163	57.6%
Jonathan Toews	CHI	2,390	1,762	57.6%

PLAYER	TEAM	WON	LOST	FO%
Joe Thornton	SJS	1,581	1,174	57.4%
Boyd Gordon	ARI/EDM	1,976	1,523	56.5%
Jerred Smithson	FLA/EDM/TOR	391	303	56.3%
Antoine Vermette	ARI/CHI	2,359	1,840	56.2%
Matt Cullen	MIN/NSH	838	669	55.6%
Paul Stastny	COL/STL	1,735	1,414	55.1%
Joe Pavelski	SJS	1,659	1,354	55.1%
Travis Zajac	NJD	1,908	1,567	54.9%
Ryan Kesler	VAN/ANA	1,851	1,522	54.9%
Steve Ott	BUF/STL	952	784	54.8%
Mikko Koivu	MIN	2,232	1,841	54.8%
Claude Giroux	PHI	2,638	2,182	54.7%
Jussi Jokinen	CAR/PIT/FLA	536	445	54.6%
Dominic Moore	NYR	939	783	54.5%
Jay Beagle	WSH	762	639	54.4%
Jay Vitale	PIT/ARI	711	599	54.3%
Jason Spezza	OTT/DAL	1525	1292	54.1%
Jay McClement	TOR/CAR	1,428	1,215	54.0%
Jeff Halpern	NYR/MTL/ARI	504	429	54.0%
Pavel Datsyuk	DET	1,535	1,309	54.0%
Derek MacKenzie	CBJ/FLA	991	851	53.8%
Jordan Staal	CAR	1,760	1,516	53.7%
Jarret Stoll	LAK	1,688	1,458	53.7%
Matt Hendricks	WSH/NSH/EDM	279	241	53.7%
Andrew Desjardins	SJS/CHI	568	491	53.6%
Marcel Goc	FLA/PIT/STL	1,440	1,264	53.3%
Brandon Dubinsky	CBJ	1,278	1,127	53.1%
Ryan O'Reilly	COL	1,146	1,015	53.0%
Anze Kopitar	LAK	1,998	1,771	53.0%
Martin Hanzal	ARI	1,243	1,105	52.9%
David Backes	STL	1,715	1,528	52.9%
Marcus Kruger	CHI	1,023	913	52.8%
Troy Brouwer	WSH	585	524	52.8%
David Krejci	BOS	1,295	1,161	52.7%
Valtteri Filppula	DET/TBL	1493	1,341	52.7%

PLAYER	TEAM	WON	LOST	FO%
Brian Boyle	NYR/TBL	981	883	52.6%
Jeff Carter	LAK	1,165	1,056	52.5%
Brad Richardson	LAK/VAN	836	764	52.3%
Matt Duchene	COL	1,655	1,513	52.2%
Ryan Johansen	CBJ	1,815	1,663	52.2%
Eric Staal	CAR	1,624	1,489	52.2%

Note: Only players who participated in a minimum of 500 faceoffs are included.

[* Raw faceoff data obtained from the NHL's official website, accessed April 9, 2015, http://www.nhl.com.]

To me, what's most striking about this leaderboard is the number of readily available replacement-level players. Even near the top of the list, there are several centres who are paid near the league minimum and have changed teams and/or been on waivers frequently over the past few years.

How important is faceoff success perceived to be when the league's best options are considered so disposable? Either these are awful players overall who are begrudgingly given NHL jobs because they can do this one thing extremely well, or there's a universal perception that faceoff talent is of very limited value.

The drop-off in faceoff percentages as the list progresses is also quite striking, meaning that even the scarcity of this talent isn't enough to raise its perceived value. There are barely a dozen players who win more than 55% of their faceoffs and only a few more who average even a single extra faceoff win per game—how important can that be?

The True Value of Winning a Faceoff

Alan Ryder's first law of hockey analytics is that "winning is what matters."* Ultimately, nothing in our field truly matters unless it can somehow be measured in terms of wins and losses. Ryder's second law is that "goals for and against are the *only* factors that affect winning," which basically means that statistics should be measured in terms of goals wherever and however possible.

[* Alan Ryder, "The Ten Laws of Hockey Analytics," Hockey Analytics, January 2008, accessed April 9, 2015, http://hockeyanalytics.com/2008/01/the-ten-laws-of-hockey-analytics/.]

How can faceoffs be measured in terms of goals? First of all, we can count how many goals were scored within 20 seconds of each faceoff win. Michael

Schuckers led a group that did just that, also breaking down the data by zone and by manpower situation.* Obviously, faceoff wins in the neutral zone don't lead to a lot of scoring, but having the man advantage sure does.

[* Michael Schuckers, Tom Pasquali, and Jim Curro, "An Analysis of NHL Faceoffs" (St. Lawrence University and Statistical Sports Consulting, 2012), accessed April 9, 2015, http://statsportsconsulting.com/main/wp-content/uploads/FaceoffAnalysis12-12.pdf.]

NUMBER OF FACEOFF WINS TO SCORE/PREVENT ONE GOAL*

	ALL ZONES	OFFENSIVE ZONE	NEUTRAL ZONE
All situations	76.5	60.1	163.8
Even strength	101.6	80.2	170.4
Special teams	40.9	35.4	128.6

[* Faceoff goal data from Mike Schuckers et al., http://statsportsconsulting.com/main/wp-content/uploads/FaceoffAnalysis12-12.pdf.]

These results start to paint at least a high-level picture of the value of a strong faceoff specialist. Since every 76.5 faceoffs won are worth an extra goal, on average, a top player like Patrice Bergeron has been worth up to 12 goals over the past three seasons. Of course, a lot of that depends on the zones in which he's used and how much special-teams time he's assigned.

Then again, that calculation assumes that if Bergeron wasn't available, then the next available centre would win half his draws, which probably isn't true. Some of his faceoffs would go to a usable option like David Krejci, but the remainder would likely be assigned to someone with a faceoff percentage below 50%. That means that Bergeron is likely worth far more than those 12 goals. But how much more?

What Is Replacement Level?

To calculate the value of any aspect of the game, the first step is to figure out the minimum level at which any readily available NHLer would perform. In the case of faceoffs, that's certainly lower than 50%, but by how much?

In his aforementioned study, Schuckers defined replacement level as any player who took fewer than 20 draws over the course of the three seasons he studied. In his catch-all WAR statistic, Andrew Thomas chose a model that was based on 50 draws a season.* Wherever the exact line is drawn, this results in a centre who wins faceoffs roughly 44% of the time. That means that Bergeron has actually been worth approximately 19 goals over these three seasons—12 goals above 50% and seven more goals above a 44% replacement.

[* Andrew Thomas, "The Road to WAR, Part 7: What Do We Mean by "Replacement"? A Case Study with Faceoffs," *WAR on Ice* (blog), March 23, 2015, http:

//blog.war-on-ice.com/the-road-to-war-part-7what-do-we-mean-by-replacement-a-case-study-with-faceoffs/.]

The problem with this definition of replacement level is that it could result in some selection bias, since only those who were particularly weak and/or inexperienced in the faceoff circle would be restricted to between 20 and 50 opportunities or less. Anyone who demonstrated any aptitude whatsoever would soon play himself out of that group, whether he was a replacement-level option or not.

That's why Matt Cane defined replacement level differently, judging it based on a player's share of all the faceoffs that occurred while he was on the ice.* Obviously players like Bergeron handle virtually every on-ice draw while others handle fewer, depending on the skill difference between them and their linemates. Consequently, Cane found a pretty direct relationship between a player's share of faceoffs and his faceoff-winning percentage.

[* Matt Cane, "Calculating Replacement Level Faceoff Win Percentage," *Puck Plus Plus* (blog), March 8, 2015, http://puckplusplus.com/2015/03/08/calculating-replacement-level-for-faceoff-win-percentage/.]

Selecting players who handle at least 65% of on-ice draws as the cut-off, he found that so-called "everyday centres" win 50.8% of their draws, while the replacement-level choices win 47.4%. So Cane would probably peg Bergeron's success as being worth roughly 14 or 15 goals over these three seasons, relative to replacement level.

Net Goals Post Faceoff (NGPF)

So far we have dealt only with estimates. Whether we land on 12, 15, or 19, that's not Boston's *actual* goal differential after Bergeron's draws, just an estimate based on his faceoff percentages and some league averages.

Even Bergeron's faceoff-winning percentage itself is somewhat of an estimate, since it's based on a scorekeeper's subjective opinion of who won each draw. For all we know, he was good at chipping the puck onto a teammate's stick but not in the kind of clean fashion that immediately leads to scoring opportunities. To find the league's best faceoff specialist, we need an approach that bases its measurements on what actually happens after each faceoff, and not on a scorekeeper's judgment or on estimates based on league-wide numbers.

That's why the new way to measure faceoff success is not to subjectively count wins and losses based on who touched the puck first but to count up the events that occur after the draw. That's exactly what Craig Tabita had in mind when he introduced net goals post faceoff (NGPF), which is simply the team's goal differential within 10 seconds of the player's faceoffs.*

[* Craig Tabita, "Redefining Faceoff Success Using Shot Data," *Hockey Prospectus* (blog), January 31, 2015, http://www.hockeyprospectus.com/redefining-faceoff-success-using-shot-data/.]

Why 10 seconds? Because, according to his data in the following chart, that's roughly when the sizable advantage enjoyed by the offensive team (in dark grey) over the defensive team (in light grey) begins to disappear.

Selecting 10 seconds instead of Schuckers's choice of 20 seconds allows Tabita to really focus on those goals that occurred moreso as a direct result of faceoff success. However, it doesn't allow for the inclusion of neutral zone faceoffs, since those scoring opportunities will take a little bit longer to develop and require a lot more than faceoff success to achieve. That is, a player would have to win the draw, gain the zone, keep possession, and translate that advantage into an attempt on goal. At this point, the benefit of winning a faceoff starts to get watered down in the mix anyway.

Interestingly, leaving out neutral zone faceoffs still results in a league-wide estimate of 78.7 faceoff wins per goal (at even strength), which is almost exactly what Schuckers found (80.2). So even if it is an estimate, it's one in which we can have confidence.

Getting back to Bergeron, the Bruins scored exactly 16 even-strength goals (GF) off of his non-neutral zone draws over the three seasons in question while allowing seven goals against (GA) for an NGPF of +9 goals. Given that he takes a lot more draws in the defensive zone than in the offensive zone, a league-average centre would have actually posted a NGPF of −1.4 (exp).

All of this means that Bergeron's adjusted NGPF is +10.4 extra goals relative to a league-average centre, which is second in the league over this time span. That may seem low, but remember that this is relative to a league-average centre, not a replacement-level centre. It's also at even strength only and ignores neutral zone faceoffs. In reality, his true value likely remains in the previously calculated 15-goal range, possibly even higher.

TOP 20 FACEOFF SPECIALISTS, 2012–13 TO 2014–15*

PLAYER	TEAM	OFF ZONE	DEF ZONE	GF	GA	NGPF	EXP	ADJ NGPF
David Desharnais	MTL	899	596	19	6	13	1.9	11.1
Patrice Bergeron	BOS	1,000	1,219	16	7	9	−1.4	10.4
Sidney Crosby	PIT	1,092	931	20	9	11	1.0	10.0
Mikael Granlund	MIN	598	493	13	3	10	0.7	9.3
Ryan Getzlaf	ANA	864	916	14	6	8	−0.3	8.3
Jamie Benn	DAL	634	420	11	3	8	1.4	6.6
Bryan Little	WPG	888	909	14	8	6	0.1	6.1
Mathieu Perreault	WSH/ ANA/ WPG	442	429	6	0	6	−0.1	5.9
Paul Gaustad	NSH	306	1,233	5	5	0	−5.9	5.9
Nicklas Backstrom	WSH	1,098	925	15	8	7	1.1	5.9
David Backes	STL	663	843	6	2	4	−1.1	5.1
Nate Thompson	TBL/ ANA	567	744	7	3	4	−1.1	5.1
Jay McClement	TOR/ CAR	455	924	6	4	2	−3.0	5.0
Ryan White	MTL/ PHI	138	246	5	1	4	−0.7	4.7
Antoine Vermette	ARI/ CHI	991	1,018	10	6	4	−0.2	4.2
Zack Smith	OTT	519	657	7	4	3	−0.9	3.9
Anze Kopitar	LAK	956	848	9	5	4	0.7	3.3

PLAYER	TEAM	OFF ZONE	DEF ZONE	GF	GA	NGPF	EXP	ADJ NGPF
Nick Bjugstad	FLA	565	567	6	3	3	0	3.0
Lars Eller	MTL	542	690	6	4	2	−0.9	2.9
Jonathan Toews	CHI	1,102	778	11	6	5	2.1	2.9

[* Faceoff data obtained from Craig Tabita, faceoffs.net, accessed April 9, 2015, http://www.faceoffs.net.]

While this precise count of goals scored after faceoff wins is nice to have, 16 data points over three years isn't nearly a large enough sample size on which to base any kind of evaluation. Theoretically, a mediocre centre who got a few lucky bounces, faced some weak opposing goaltenders, and/or played alongside some particularly effective linemates could generate the same number of extra goals as an elite option like Bergeron.

For example, take David Desharnais, the one player who fared even better than Bergeron on that leaderboard but doesn't appear among the faceoff percentage leaders. Desharnais won only 51.4% of his faceoffs over the past three seasons, but he has helped Montreal score 19 even-strength goals off the draw, which is tied with Sidney Crosby for the league lead. He has a raw NGPF of +13, which is still a mighty +11.1 once adjusted for his larger volume of offensive zone draws.

Either Desharnais is winning fewer draws but much more cleanly, or he and his linemates have been on the right side of the bounces. That's the universal drawback of using goal-based statistics when evaluating individual players; it's just a lot harder to separate the good from the lucky.

Net Shots Post Faceoff (NSPF)

One of the problems with using goals, either in this situation or in any other statistical analysis, is that their relative scarcity leaves them vulnerable to the effect of random variation, among so many other factors. That's why Tabita generally relies on his net shots post faceoff (NSPF) metric instead, which is the balance of unblocked attempted shots for (SAF) minus those against (SAA) within 10 seconds of a faceoff. Again, roughly 10 seconds after the draw is when the advantage enjoyed by the offensive team (in dark grey) begins to disappear.

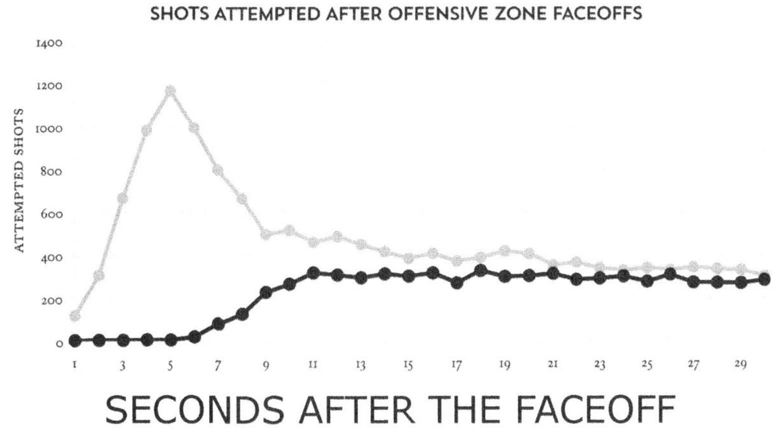

With this new view incorporated into the table that follows, the Canadiens have actually given up about 20 *more* shot attempts with Desharnais than with a league-average centre, while the Bruins have enjoyed an extra 88 shot attempts thanks to Bergeron. Over this three-year span, that places Boston's superstar second in the entire NHL, just three shot attempts behind Philadelphia's Claude Giroux.

Just as it was in the goal-based analysis, this adjusted form of NSPF is calculated by taking the player's raw NSPF and subtracting what a league-average centre would be expected to achieve (exp) with the same number of draws in each zone.

TOP 50 FACEOFF SPECIALISTS, 2012–13 TO 2014–15*

PLAYER	TEAM	OFF ZONE	DEF ZONE	SAF	SAA	NSPF	EXP	ADJ NSPF
Claude Giroux	PHI	1,253	1,124	272	162	110	18.8	91.2
Patrice Bergeron	BOS	1,000	1,219	196	140	56	−31.9	87.9
Joe Pavelski	SJS	791	775	167	101	66	2.3	63.7
Paul Gaustad	NSH	306	1,233	77	155	−78	−135.1	57.1
David Backes	STL	663	843	152	122	30	−26.2	56.2
Boyd Gordon	ARI/ EDM	407	1,414	95	189	−94	−146.8	52.8
Nicklas Backstrom	WSH	1,098	925	246	168	78	25.2	52.8

PLAYER	TEAM	OFF ZONE	DEF ZONE	SAF	SAA	NSPF	EXP	ADJ NSPF
Ryan Johansen	CBJ	929	896	159	117	42	−9.8	51.8
Mikko Koivu	MIN	1,092	786	208	113	95	44.6	50.4
Joe Thornton	SJS	693	744	142	107	35	−7.4	42.4
Anze Kopitar	LAK	956	848	198	140	58	15.7	42.3
Vladimir Sobotka	STL	357	354	77	42	35	0.4	34.6
David Krejci	BOS	829	556	159	87	72	39.8	32.2
Henrik Sedin	VAN	1,072	702	188	102	86	53.9	32.1
Sidney Crosby	PIT	1,092	931	229	175	54	23.5	30.5
Mike Richards	LAK	597	482	126	81	45	16.8	28.2
Vernon Fiddler	DAL	532	771	101	108	−7	−34.8	27.8
Casey Cizikas	NYI	477	551	89	72	17	−10.8	27.8
Mike Fisher	NSH	720	700	137	107	30	2.9	27.1
Zenon Konopka	MIN/ BUF	115	376	27	38	−11	−38.0	27.0
Jordan Staal	CAR	862	846	183	155	28	2.3	25.7
Dominic Moore	NYR	237	665	52	90	−38	−62.4	24.4
Patrick Marleau	SJS	203	186	52	26	26	2.5	23.5
Jeff Carter	LAK	633	509	133	92	41	18.1	22.9
Brad Richardson	LAK/ VAN	366	554	68	74	−6	−27.4	21.4
Jim Slater	WPG	222	360	58	57	1	−20.1	21.1
Ryan Kesler	VAN/ ANA	788	818	137	121	16	−4.4	20.4
Darren Helm	DET	265	219	57	30	27	6.7	20.3
Jason Spezza	OTT/ DAL	841	615	162	109	53	32.9	20.1
Calle Jarnkrok	NSH	215	166	45	18	27	7.1	19.9
Rich Peverley	BOS/ DAL	259	236	63	40	23	3.4	19.6
Tyler Johnson	TBL	613	534	114	85	29	11.5	17.5
Jamie Benn	DAL	634	420	129	82	47	31.2	15.8
Joakim Andersson	DET	352	349	58	42	16	0.4	15.6
Brian Boyle	NYR/ TBL	277	563	55	82	−27	−41.7	14.7
Frans Nielsen	NYI	575	675	88	88	0	−14.6	14.6
Sean Monahan	CGY	775	790	125	115	10	−2.2	12.2

PLAYER	TEAM	OFF ZONE	DEF ZONE	SAF	SAA	NSPF	EXP	ADJ NSPF
Logan Couture	SJS	622	685	112	109	3	−9.2	12.2
Ryan Strome	NYI	207	208	44	32	12	−0.1	12.1
Nick Bonino	ANA/ VAN	645	721	107	106	1	−11.1	12.1
Michael Latta	WSH	132	138	26	15	11	−0.9	11.9
Nate Thompson	TBL/ ANA	567	744	104	118	−14	−25.8	11.8
John Tavares	NYI	1,121	645	190	109	81	69.4	11.6
Chris Mueller	NSH/ DAL/ NYR	70	101	19	12	7	−4.5	11.5
Mark Arcobello	EDM/ NSH/ PIT/ ARI	445	313	79	49	30	19.2	10.8
Bo Horvat	VAN	226	257	48	42	6	−4.5	10.5
Jay McClement	TOR/ CAR	455	924	94	152	−58	−68.4	10.4
Mathieu Perreault	WSH/ ANA/ WPG	442	429	82	70	12	1.9	10.1
Ryan White	MTL/ PHI	138	246	22	28	−6	−15.7	9.7
Eric Fehr	WSH	311	425	63	70	−7	−16.6	9.6

[* Faceoff data obtained from Craig Tabita, faceoffs.net, accessed April 9, 2015, http://www.faceoffs.net.]

To convert these shot attempts into goals, divide by 20. Offensively, 4.7% of these unblocked shot attempts resulted in goals, and when a faceoff was lost and the defensive players skated up the ice, 8.5% of those attempted shots resulted in goals, for an average of just under 5% overall (since the former are more common).

From that perspective, Giroux's success at even strength should have been worth 4.5 goals more than a league-average centre over the past three seasons, but the Flyers have both scored and allowed 12 goals following his draws. That means that Giroux is giving his team many more scoring chances but that Philadelphia has been allowing the opponents to convert a greater portion of theirs.

As for Bergeron, has his success at even strength over the past three years been worth 10.4 goals (relative to league average), as calculated by looking strictly at goals, or 4.3 goals, as calculated based on shot attempts? Unlike Giroux, Bergeron has been playing in front of a Vezina-winning goalie and a Norris finalist on defence, who have denied opponents the opportunity to capitalize on his rare faceoff losses. If he played on a defensively vulnerable team like Philadelphia, then it's possible that his goal-based numbers would be as modest as Giroux's. Reducing the impact of team factors is yet another reason why shot-based data is frequently preferred.

One final reason that shot-based data is so popular in the world of hockey analytics is its close relationship with puck possession. All things being equal, the larger a team's share of all on-ice attempted shots, then the more that team was in possession of the puck and/or in the offensive zone. Given that a faceoff is basically all about gaining possession of the puck, it would make sense if there was some kind of relationship between the two. And sure enough, two separate studies, at *Hockey Wilderness* and *Hockey Graphs,* have found that there's a minor but real relationship.* Specifically, variance in faceoff success explains between 17% and 21% of the variance in team possession numbers. Essentially, faceoffs are one of the many types of battles that do indeed contribute to a team's puck possession numbers.

[* Mntrumpterguy, "Connecting the Dots: Faceoffs and Possession," *Hockey Wilderness* (blog), November 6, 2014, http://www.hockeywilderness.com/2014/11/6/7155985/faceoffs-nhl-possession-correlation-do-faceoffs matter; Garret Hohl, "The Relationship Between Corsi% and Winning Faceoffs," *Hockey Graphs* (blog), January 15, 2015, http://hockey-graphs.com/2015/01/15/the-relationship-between-corsi-and-winning-faceoffs/.]

Factors Impacting Faceoff Success

Regardless of whether faceoff success is measured with a subjectively judged win/loss percentage or with more recent methods that are based on which team attempts the next shot, there are a number of factors that can affect the results. Keeping outside factors in mind, and accounting for them whenever possible, is what makes the difference between a really useful statistical perspective and one that's just a little bit off.

For example, Paul Gaustad and Joe Thornton have nearly identical win/loss totals over the past three seasons combined and similar shot-based results. Since Gaustad's record was 1,583/1,163, Thornton's was 1,581/1,174, and their adjusted NSPF is only 15 attempted shots apart, it's easy to walk away with the impression that they're quite comparable. But are they really?

- Gaustad takes four times more even-strength faceoffs in the defensive zone than in the offensive zone, while Thornton takes roughly the same number in each location.
- Nashville's giant has taken 558 faceoffs short-handed and only 46 with the man advantage, while San Jose's captain has taken only 192 when down a man and 406 on the power play.
- Gaustad's most frequent linemates have been Eric Nystrom, Gabriel Bourque, and Rich Clune, who aren't exactly the type of players who generate post-win shot attempts. In contrast, Thornton is typically out there with skilled linemates like Joe Pavelski and Tomas Hertl.

Despite the similarity of their numbers, one of them is clearly a better candidate for the league's best faceoff specialist than the other.

Right now, it appears that Bergeron is the man to beat, both by reputation and based on the underlying numbers. From 2012–13 through 2014–15 combined, he was one of only two players to have won at least 60% of his draws, posted a second-best adjusted NGPF of 10.4 goals, and had a second-best adjusted NSPF of 87.9. All of these results are over a second-highest 4,567 faceoffs, which were taken in all three zones and in all manpower situations.

Before we fit Boston's faceoff ace for his championship belt, let's consider if he has received an unfair advantage by taking a closer look at all the factors that can influence these results, including the impact of random variation, the quality of opponent, the faceoff's location, the manpower situation, home-ice advantage, and more.

1. RANDOM VARIATION

Zenon Konopka is the only player with a higher faceoff-winning percentage than Bergeron over these three seasons, and yet he's rarely heralded as the league's best because he has taken just 834 draws to Bergeron's 4,567.

Every player has a true faceoff talent, which requires reverse-engineering the results to figure out. Obviously, 4,567 faceoffs over three years is enough to feel pretty confident that Bergeron's true talent is quite close to the observed 60%. In fact, statistically, we know with 95% confidence that his true talent falls somewhere between 58.5% and 61.4%.

The picture is not nearly as clear with Konopka, however. The confidence interval itself is easy enough to calculate because it is simply a player's faceoff-winning percentage plus or minus 1.96 times the standard error. The problem with Konopka is that his far lower sample of data means that his standard error is 1.7%, which is much higher than Bergeron's 0.7%. Consequently, his true faceoff talent is anywhere from 57.5% to 64.1%. That's great, but it's impossible to statistically establish if he's the best. On the plus side for Konopka, he is one of only 10 centres whose confidence interval overlaps with Bergeron's.

Statistically, we don't have the information to narrow down the list any further. Everything we observe in hockey, including faceoff success, is a combination of both talent and random variation. Talent is the component that remains relatively constant throughout a player's career, especially past a certain age, while random variation is what fluctuates from season to season to one extent or another. It's what makes it so difficult to measure a player's true talent.

To estimate the extent to which random variation affects a player's faceoff-winning percentage, I calculated the (R-Squared) season-to-season relationship for every player who took at least 500 draws in back-to-back seasons from 2008–09 through 2014–15. The resulting score of 0.58 (out of 1) means that a player's faceoff percentage is mostly a result of skill but still involves a fair deal of random variation. Indeed, random variation alone accounts for up to a quarter of a player's results (one minus the square root of 0.58).

While random variation is without question the single largest outside factor that can impact a player's faceoff-winning percentage, there are several others, each of which can be calculated and adjusted for.

2. QUALITY OF COMPETITION

It stands to reason that the quality of one's opponent could have an impact on a player's success in the faceoff circle. Obviously it's a lot harder to win a draw against Bergeron than it is on a road trip through Alberta. The only real question is whether the average quality of one's opponents usually works out to be roughly a wash by season's end.

To my knowledge, this concept was first explored by Timo Seppa as part of his ultimate faceoff statistic (UFO%) in 2011, when he observed that "teams tend to send out their best faceoff men in special teams situations and for offensive and defensive zone draws. As a consequence, strong faceoff men often face other strong faceoff men, making both look mediocre."[10]

Despite the rather obvious and clear-cut nature of this concept, and the availability of the raw data in the NHL's game files, quality of faceoff competition hasn't been included in any of the so-called fancy stats websites out there. At least not until Craig Tabita launched his own in the 2014–15 season, defining quality of competition as "a weighted average of a

10 Timo Seppa, "No Advanced Faceoff Metric?," in Hockey Prospectus 2011–12, by the authors of Hockey Prospectus (CreateSpace Independent Publishing Platform, 2011), 390–394.

player's faceoff opponents' faceoff percentages against the rest of the league."*

[* Craig Tabita, "Beyond Faceoff Percentage, Part 1: Quality of Competition, faceoffs.net, November 7,2014, accessed April 9, 2015, http://faceoffs.net/blog/2014/11/faceoff-percentage-quality-of-competition/.]

The good news is that Tabita has published historical data all the way back to 2008-09, but the bad news is that there really isn't much of a difference from one player to another.

TOP 20 QUALITY OF FACEOFF COMPETITION, 2012-13 TO 2014-15*

PLAYER	TEAM	QOC
Jamie Benn	DAL	50.8%
Jonathan Toews	CHI	50.8%
Anze Kopitar	LAK	50.6%

PLAYER	TEAM	QOC
Jason Spezza	OTT/DAL	50.5%
Mikko Koivu	MIN	50.5%
Mike Fisher	NSH	50.5%
John Tavares	NYI	50.5%
Ryan Getzlaf	ANA	50.5%
Bryan Little	WPG	50.5%
David Backes	STL	50.5%
Pavel Datsyuk	DET	50.5%
Cody Hodgson	BUF	50.5%
Tyler Seguin	BOS/DAL	50.5%
Kyle Okposo	NYI	50.5%
Antoine Vermette	ARI/CHI	50.5%
Vincent Lecavalier	TBL/PHI	50.4%
Jeff Carter	LAK	50.4%
Steve Ott	BUF/ST	50.4%
Sean Monahan	CGY	50.4%
Johan Larsson	MIN/BUF	50.4%

Note: Only players with a minimum of 500 faceoffs are included.

[* Faceoff data obtained from Craig Tabita, faceoffs.net, accessed April 9, 2015, http://www.faceoffs.net.]

Over the past three seasons, only three centres in the entire NHL have faced opponents with a winning percentage higher than a mere 0.5% above league average. Even this modest result could easily be more of a consequence of random variation than actually getting dealt a tough hand.

Quality of competition is a far more important factor when comparing players across leagues. For example, consider the KHL's faceoff master, Jarkko Immonen. The former New York Ranger has won 64.7% of his

3,180 KHL faceoffs over the same three seasons. If he were playing in the NHL, would his performance rival Bergeron's?

On one hand, Immonen did win 56.6% of his 159 NHL faceoffs in 2005–06 and 2006–07. On the other hand, he won only 48.1% of his 54 draws in the 2014 Olympics, compared with 63.3% for Bergeron. Furthermore, his closest rival in the Russian faceoff circle, Jori Lehtera, won 58.9% of his 2,216 KHL draws but won just 48.7% in the Olympics and then just 51.2% for the St. Louis Blues in 2014–15. The lower quality of competition could be an important factor in Immonen's (and Lehtera's) success in the KHL.

Similarly, Sweden's faceoff king over the past three seasons is Chris Abbott, who won 60.8% of his 3,093 SHL draws, followed by Joel Lundqvist, who won 58.4% of 2,748. Abbott has never played in either the NHL or the Olympics, but Joel Lundqvist, who is the identical twin of the Rangers' star goalie Henrik Lundqvist, played 134 games over three seasons for the NHL's Dallas Stars from 2006–07 to 2008–09. During that time, Lundqvist won just 49.4% of his 391 faceoffs. Again, Abbott and Lundqvist are likely facing easier opponents in Sweden than they would compete against in the NHL.

One interesting avenue for further exploration is the effect of an opponent's handedness on faceoff success. In baseball, pitchers have an advantage when facing batters who hit with the same hand with which they throw, but no such overall pattern can be found in hockey's faceoffs. However, Tabita has found individual cases where the results can't be explained by random variation, like Ryan Getzlaf's success against right-handers or Mike Ribeiro's struggles against the same.* It's too bad for Getzlaf that roughly 63% of NHL faceoffs are taken by centres with left-handed shots.

[* Craig Tabita, "Introducing Faceoff Percentage Opponent Hand Splits, and Revealing Hidden Biases," faceoffs.net, December 16, 2014, accessed April 9, 2015, http://faceoffs.net/blog/2014/12/introducingfaceoff-percentage-opponent-hand-splits-and-revealing-hidden-biases/.]

3. HOME-ICE ADVANTAGE

Rule 76.4 in the official NHL rule book states that "the sticks of both players facing off shall have the blade on the ice, within the designated white area. The visiting player shall place his stick within the designated white area first followed immediately by the home player."*

[* National Hockey League, *National Hockey League Official Rules 2014–15,* NHL, http://www.nhl.com/nhl/en/v3/ext/rules/2014-2015-rulebook.pdf.]

This built-in rules advantage helps the home team, especially in critical draws in the defensive zone or on special teams. That's why Schuckers found that "being the home team was a significant predictor of faceoff wins" and that "being at home [increases] the win percentage by 1.5%."* Based on the eight-year chart below, that's a very accurate and consistent estimate.**

[* Michael Schuckers, Tom Pasquali, and Jim Curro, "An Analysis of NHL Faceoffs" (St. Lawrence University and Statistical Sports Consulting, 2012), April 9, 2015, http://statsportsconsulting.com/main/wp-content/uploads/FaceoffAnalysis12-12.pdf.]

[** Faceoff data obtained from Craig Tabita, faceoffs.net, accessed April 9, 2015, http://www.faceoffs.net.]

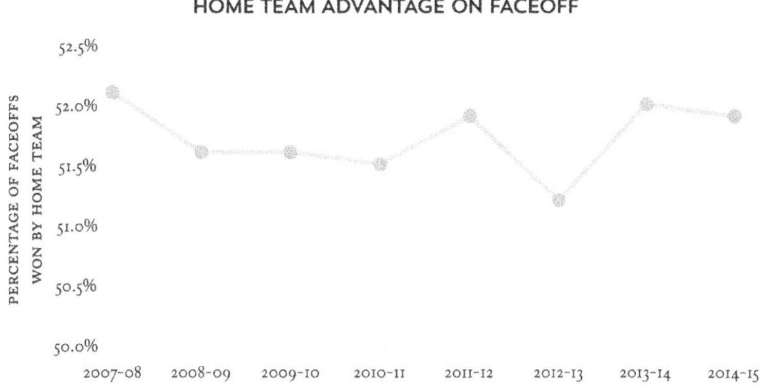

In general, there are two ways to account for factors like home-ice advantage in faceoffs:

1. break up the data based on that factor and examine each situation separately, or

2. calculate the impact of the outside factor and create an adjusted version of the statistic in question.

Seppa did the former with UFO%, limiting the calculation to road games only and only in even-strength situations. Ultimately, only 35% of faceoffs are included in his metric, which is still enough to remain useful when examining three-year periods or for each team's top few options in single-season samples.

Schuckers went the latter direction, calculating an adjusted form of faceoff percentage that took not just home-ice advantage but also manpower situation and the location of the faceoff into account. However, he found that this adjusted version of faceoff-winning percentage had a near-perfect correlation with the raw faceoff percentage. Why? Despite the obvious benefit of getting to place the stick on the ice last, the year-end impact of home-ice advantage is quite minimal, because virtually all players take roughly the same number of draws at home as they do on the road.

As of the 2015–16 season, this advantage has disappeared because it is the centre in the offensive zone who gets to place his stick down last. Personally, I wonder how much of this observed home-ice advantage is from placing the stick on the ice last and how much is due to scorekeeper bias. Given the already subjective nature of determining who won and lost a faceoff, it would be only natural if the home team was awarded a few extra wins per game. After all, we see the same pattern in statistics like shots, blocks, hits, and take-aways, so why should faceoffs be different?

Schuckers himself found that roughly 5% of all faceoffs are recorded incorrectly, and he points to certain cities as being worse than others, like Long Island's Nassau Coliseum, most notably. That's one more reason to set aside home-ice advantage, in my view, and focus on more significant outside factors.

4. THE MAN ADVANTAGE

Other than random variation, playing with the man advantage is the external factor that has the greatest impact on faceoff success, and it can vary the most from player to player.

Faceoff wins are typically scored based on which team is the first to touch the puck after the draw. Having the extra player on the ice gives the power-play centre that much more opportunity to hit an open stick. Theoretically, the short-handed centre might also be more concerned about avoiding a clean win for the opponent than about winning the draw himself, which could artificially boost the power-play advantage even further.

Initially, Schuckers calculated that being on the man advantage improves the probability of winning a faceoff by 5.7%, but recent data suggest that this has been trending down since then.* Surprisingly, the power-play boost seems to be of benefit only under the traditional perspective of faceoff wins and losses and doesn't really affect who takes the next shot.

[* Faceoff data obtained from Craig Tabita, faceoffs.net, accessed April 9, 2015, http://www.faceoffs.net.]

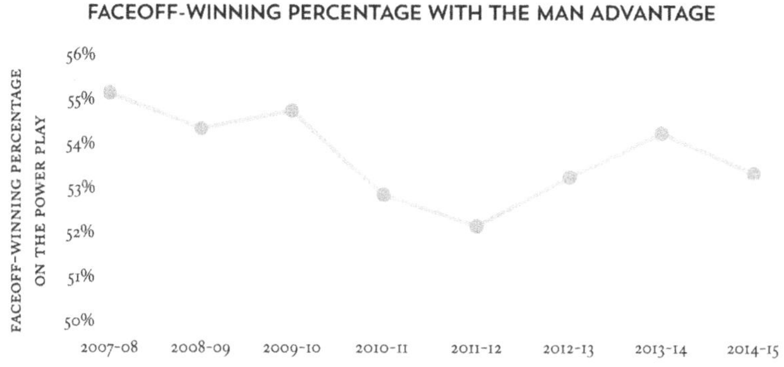

Years before Tabita, it was Gabriel Desjardins who was the first to study the value of an offensive zone faceoff win by comparing shot rates immediately after a win to shot rates immediately after a loss. He found that shot rates after wins are actually higher at even strength than on the power play.* It's up for further analysis whether this is because a team with a man advantage will take some time to set up a play for a better shot after a faceoff win or because teams killing penalties are more likely to block shots. Or for some other reason entirely.

[* Gabriel Desjardins, "Impact of Winning an Offensive Zone Faceoff, Even-Strength vs Power Play," *Arctic Ice Hockey* (blog), October 18, 2011, http://www.arc

ticicehockey.com/2011/10/18/2491154/impact-ofwinning-an-offensive-zone-faceoff-even-strength-vs-power.]

5. FACEOFF LOCATION

Naturally, the zone in which a faceoff occurs has a bearing on who gets the next shot attempt, which is why such statistics are divided by zone, but does it affect the traditional view of wins and losses?

For the last five seasons, teams in the defensive zone have had a slight advantage, but not one that Schuckers was prepared to describe as "statistically significant."* While a very slight upward trend appears visible to the naked eye, everything here can be explained by random variation. This may change in 2015–16 and beyond, when the offensive centre gets to place his stick down last.

[* Faceoff data obtained from Craig Tabita, faceoffs.net, accessed April 9, 2015, http://www.faceoffs.net.]

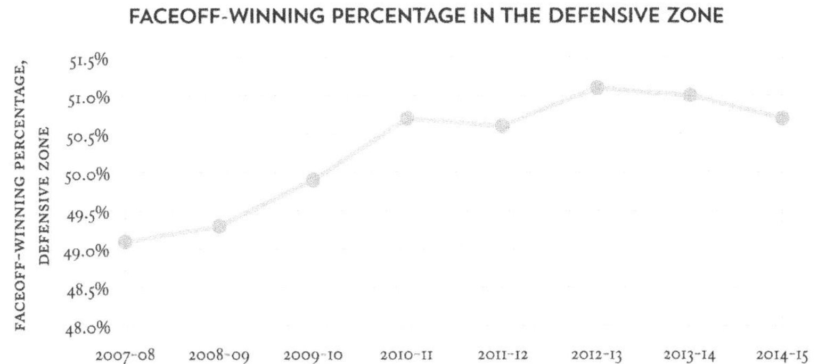

6. LINEMATES

Since faceoff wins and losses are determined by who is the first player to touch the puck, can faceoff-winning percentages be impacted by a player's linemates? After all, there must be some wingers and defencemen who are better at tying up their opponent and/or getting to the puck than others.

Matt Pfeffer studied this using a with or without you (WOWY) analysis, comparing how centres fared in the faceoff circle with and without certain players alongside them. Pfeffer managed to find some players alongside whom centres had more success winning faceoffs than with others, but rarely to the extent that went beyond the impact of random variation.*

[* Matt Pfeffer, "Can Players Impact Their On-Ice Faceoff Percentage?," Progressive Hockey, September 25, 2014, accessed April 9, 2015, http://www.progressivehockey.com/2014/09/can-players-impact-their-onice-faceoff.html.]

One notable exception was Corey Perry, with whom Anaheim centres have had a much better success rate. As a big power forward, it makes sense that Perry could be the type of player who would prove useful at assisting with faceoff wins by scrambling quickly to the right area and/or tying up his opponent from doing the same. Having Perry on his right side could also explain the earlier observation that

Anaheim's Ryan Getzlaf has far more success against right-shooting centres.

In terms of broad, general rules, Pfeffer also found that "in the offensive zone, wingers seem to have a larger potential for impact and conversely in the defensive zone defencemen seem to be able to have a larger average deviation from the mean."* Yes, I would certainly imagine that it helps Bergeron to have the 6-foot-9 all-star veteran Zdeno Chara behind him on defensive zone draws.

[* Matt Pfeffer, "Can Players Impact Their On-Ice Faceoff Percentage?," Progressive Hockey, September 25, 2014, accessed April 9, 2015, http://www.progressivehockey.com/2014/09/can-players-impact-their-onice-faceoff.html.]

7. AGE

Is age a factor in faceoff success? Given that the average faceoff-winning percentage increases steadily until age 26, it certainly seems that way. Taking the average results for all players at each age is a simple but incorrect way to build an age curve for any statistic, because of the skewing effect of survivorship bias (among other factors). That is, the better players will get more opportunities, which skew the results.

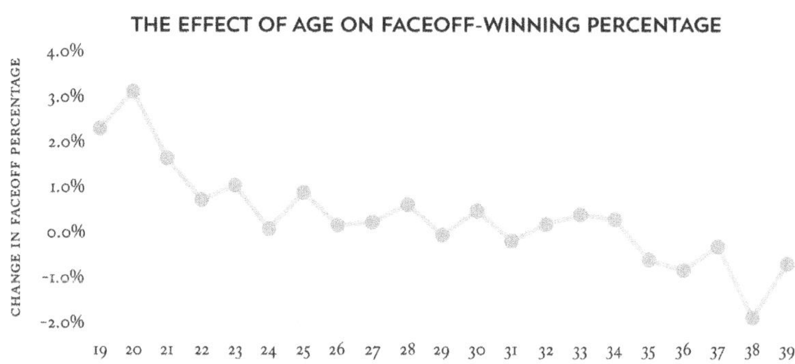

The age curve above was built correctly, in the fashion described in the team-building chapter (see pages 12–74), based on a set of matched pairs for every player who took at least 200 faceoffs in back-to-back seasons, going all the way back to the 2008–09 season.*

[* The age curve graph uses data obtained from the NHL's official website, accessed April 9, 2015, http://www.nhl.com.]

This age curve shows the average change in faceoff percentage when players go from one age to the next. For example, centres going from age 19 to 20 will improve their faceoff-winning percentage by over 3.0%, on average, while those going from 20 to 21 will improve it by about half as much.

These results prove that the conventional wisdom is correct. Players do tend to experience a rapid improvement in faceoff-winning percentage when they're young and then manage to maintain it and/or make incremental improvements throughout their careers. Even the decline that begins at age 35 is barely noticeable, and it could easily be a simple consequence of the relatively small number of remaining centres.

In essence, winning faceoffs is a skill that only gets better with age and an area where the more experienced players will usually have the edge.

Closing Thoughts

Objective statistical analysis rarely produces definitive results in a game as complex as hockey, but they can still paint a compellingly clear picture. In this case, the conventional wisdom that Patrice Bergeron is the league's best faceoff specialist appears to be valid. Studying the combined data from 2012–13 through 2014–15, very few cases can be made for any other player.

- In conventional terms, only Zenon Konopka has a higher faceoff-winning percentage over the past three years but in far fewer attempts.
- Only Philadelphia's Claude Giroux has taken more draws over the past three years, and only Giroux has generated more shot attempts off those faceoffs than Bergeron, but the Flyers' captain has below-average goal-based data.
- Similarly, only Montreal's David Desharnais has generated more goals off his victories than Boston's superstar, but he did so with a far lower winning percentage.

Nashville's Paul Gaustad joins Giroux as potentially Bergeron's strongest rival. His numbers are inferior to Bergeron's, but he has the disadvantages of taking far more draws while down a man and with inferior linemates. It's certainly possible that if Gaustad was used in a comparable fashion to Bergeron, then his numbers would match.

While all signs point to Bergeron, the results aren't nearly as fascinating as the process by which they were discovered. For example, we discovered how the traditional way of rating players using a subjectively judged winning percentage isn't nearly as useful as new statistics like NSPF, which count how many shot attempts a team achieves after each player has taken the draw. Regardless of how faceoff success is measured, we also discovered to what extent several factors can influence the results, like random variation, home-ice advantage, manpower situation, faceoff location, quality of linemates, quality and handedness of opponents, and even the player's age.

What makes this whole process so important is how this same approach can be applied to study other elements of a hockey game. Analysts have already earned front-office jobs examining manually counted breakouts, zone entries, and passing information, and once better player tracking data becomes available, the sky is the limit. Stretch passing, playing it up the boards, forcing turnovers or icings, and pretty much everything else can be studied in this same fashion.

While there aren’t many centres who have both the talent and the opportunity to have the same impact as Bergeron in the faceoff circle, this is just one of many types of puck battles. Once the player tracking data comes in, I wouldn’t be at all surprised to learn that Bergeron is having the same kind of impact everywhere else, too.

WHO IS THE BEST SHOT-BLOCKER?

by IAIN FYFFE

The NHL keeps a number of stats that are not really seriously considered by hockey analysts, despite getting a lot of play in the popular hockey media, and some insiders share this disdain. Ken Hitchcock once said, "If you're blocking shots it means the puck is in your zone a lot ... When you're a shot-blocking team, you never get it back."* This dismissiveness is not merely the result of irrational disdain for real-time scoring systems stats (such as blocked shots, hits, and takeaways) by hockey analysts and insiders. There are good reasons why the raw RTSS data should be discounted.

[* Jason Brough, "Hitch Thinks Blocked Shots Are 'The Most Useless Stat' in the NHL," *Pro Hockey Talk* (blog), NBC Sports, April 25, 2014, http://prohockeytalk.nbcsports.com/2014/04/25/hitch-thinksblocked-shots-are-the-most-useless-stat-in-the-nhl/.]

Take the blocked-shots stat, for example. Hitchcock has a point. You might think that blocking lots of shots should produce significant value for your team, since they prevent pucks from reaching your net, which makes scoring goals rather difficult for the opponents. So the shot-blocking leaderboard should be made up of some very valuable defensive players. But raw shot-blocked totals should be discounted for a very good reason: the teams that block the most shots tend to be bad teams, and the players that block the most shots tend to play for bad teams. So while high blocked-shots totals might be touted as a sign of skill, in reality it may only be a sign of playing for a bad team. Why is this? Quite simply, bad teams tend to allow their opponents to direct a lot of pucks toward goal, and with more shots going toward your net it's easier to accumulate a high count of blocked shots, even if you're blocking a relatively low proportion of them.

So it's the correlation between being a bad team and having lots of shots to block that results in the analytical disdain toward the blocked-shots stat. If lots of blocked shots are a sign of being a bad team, then surely no credit should be attributed to them with respect to winning hockey games, and the essence of hockey analytics is determining what does and does not contribute to winning hockey games.

But while this objection to the value of the blocked shots holds true on the macro level (looking just at team or player totals), on the micro level the same approach is not valid. An individual shot that gets through to the net is more likely to score than a shot that is blocked, so when a shot has already been directed at your net, blocking that shot produces value for your team in the sense of saving goals against.

So during gameplay, blocking a shot that has been directed at your net is better than not blocking it, generally speaking, and as such the blocked-shot stat should reflect some value to your hockey team, assuming you can consider the context appropriately. And since here at Hockey Abstract we are concerned about not only popularizing the analysis of hockey stats but also analyzing popular hockey stats, we're going to examine this oft-quoted but little-scrutinized statistic.

NHL BLOCKED-SHOTS LEADERS, 2012–13 TO 2014–15*

PLAYER	TEAM	POS	GP	BKS
Kris Russell	STL/CGY	D	180	486
Dan Girardi	NYR	D	209	393
Andrew MacDonald	NYI/PHI	D	188	380
Brooks Orpik	PIT/WSH	D	196	373
Roman Josi	NSH	D	201	364
T.J. Brodie	CGY	D	209	354

PLAYER	TEAM	POS	GP	BKS
John Carlson	WSH	D	212	353
Mike Weber	BUF	D	174	341
Andrei Markov	MTL	D	210	332
Shea Weber	NSH	D	205	331
Josh Gorges	MTL/BUF	D	160	330
Ron Hainsey	WPG/CAR	D	210	329
Dennis Wideman	CGY	D	172	323
Nick Schultz	EDM/CBJ/PHI	D	197	317
Brent Seabrook	CHI	D	211	314
Brenden Dillon	DAL/SJS	D	208	309
Johnny Boychuk	BOS/NYI	D	191	309
Karl Alzner	WSH	D	212	308
Alex Pietrangelo	STL	D	209	307
Jan Hejda	COL	D	205	306

[* Data obtained from the NHL's official website, accessed May 12, 2015, http://www.nhl.com.]

This table doesn't tell us very much, of course, since it doesn't consider context. Specifically, it doesn't consider ice time or manpower situation, and more importantly it doesn't consider just how many shots these players are facing, which of course determines the opportunities they have to block shots. At best, this might identify some candidates for best shot-blocker, but further analysis is required.

First, we should consider only even-strength situations. The dynamics of the play on the ice in odd-man situations might introduce bias into the results; players who kill relatively more penalties might have an advantage or disadvantage, for example. And then we need a measure of opportunity, which is the number of unblocked shot attempts taken by the other team while the player is on the ice (USATA). In this way, we can calculate the proportion of the opponent's shot attempts that the player blocked when he was on the ice. While this is still far from a perfect measure, it is a great improvement over raw shots-blocked totals.

The proportion of shots blocked should be much more informative than the raw totals, and this rate stat does not seem to suffer from the strong correlation to being on a bad team that simple blocked shots does. Unlike the raw total, there does not seem to be a strong relationship between the proportion of shots directed at the net that a team blocks and allowing a lot of goals. Some excellent defensive teams block a high proportion of shots, as do some bad defensive teams. So by translating blocked shots into a rate stat, we improve its value.

Before we look at the results, there is one final thing to consider. Whenever you are using a rate stat, one that compares the rate that one statistic occurs relative to another, you have to consider the problem of small sample sizes when ranking players. A player who plays one game and happens to block one out of the three even-strength shots the opponents take while he's on the ice will rate very highly, since his EBS% (even-strength blocked-shot percentage) would be 33.33%. Often this problem is dealt with by applying an arbitrary cut-off in order to be considered; for example, some number of games played or number of shots faced must be attained before the player will be considered and rated. But a more sophisticated approach to this problem is to apply the concept of regression to the mean.

Regression to the mean recognizes the fact that a statistic is more likely to be closer to the average on later measurements if it has an extreme result on early measurements. If a player blocks three out of the first six shots his team faces, we expect that he will not continue to block shots at such a high rate but will block shots at a more average rate as the season goes on. So instead of applying some arbitrary minimum number of games played or minutes played or USATA, we can apply an adjustment to players who faced relatively few USATA for a season. Any player who faced less than 1,000 USATA for a season essentially has additional USATA added to his record, blocked at the league-average rate for his position, to determine his final rating, which we'll call regressed even-strength blocked-shot percentage (REBS%).

Take Calvin de Haan's 2013–14 season as an example. He had an EBS% of 10.88 with 809 USATA. So we take the weighted average of this and an EBS% of 7.79 (the league average for NHL defencemen that season) on 191 USATA (the difference between 1,000 and his actual USATA of 809), which results in an REBS% of 10.29. The proper interpretation of this value is that de Haan blocked the equivalent of 10.29% of his opponent's shots taken while he was on the ice in even-strength situations.

NHL LEADERS IN REBS%, 2013–14*

PLAYER	TEAM	POS	GP	ESBK	USATA	EBS%	REBS%
Kris Russell	CGY	D	68	186	1,388	13.40	13.40
Zbynek Michalek	PHX	D	59	124	952	13.03	12.77
Andrew MacDonald	NYI/PHI	D	82	201	1,626	12.36	12.36
Chris Butler	CGY	D	82	186	1,518	12.25	12.25
Ladislav Smid	EDM/CGY	D	73	139	1,168	11.90	11.90
Raphael Diaz	MTL/VAN/ NYR	D	63	102	813	12.55	11.66
Douglas Murray	MTL	D	53	89	686	12.97	11.35
Mike Weber	BUF	D	68	128	1,161	11.02	11.02
Josh Gorges	MTL	D	66	127	1,158	10.97	10.97
Nicklas Grossmann	PHI	D	78	136	1,280	10.63	10.63
Mark Fraser	TOR/EDM	D	42	75	609	12.32	10.55
Christopher Tanev	VAN	D	64	103	984	10.47	10.42
Alexander Edler	VAN	D	63	113	1,089	10.38	10.38
Mark Stuart	WPG	D	69	115	1,113	10.33	10.33
Calvin de Haan	NYI	D	51	88	809	10.88	10.29
Nick Schultz	EDM/CBJ	D	69	104	1,018	10.22	10.22
Niklas Hjalmarsson	CHI	D	81	117	1,147	10.20	10.22
Barret Jackman	STL	D	79	99	969	10.22	10.14
John-Michael Liles	TOR/CAR	D	41	76	675	11.26	10.13
Roman Polak	STL	D	72	95	919	10.34	10.13

[* Raw data obtained from War on Ice (blog), http://www.waronice.com.]

NHL LEADERS IN REBS%, 2014–15*

PLAYER	TEAM	POS	GP	ESBK	USATA	EBS%	REBS%
Kris Russell	CGY	D	79	228	1,676	13.60	13.60
Christopher Tanev	VAN	D	70	120	991	12.11	12.07
Johnny Boychuk	NYI	D	72	119	1,029	11.56	11.56
Mike Weber	BUF	D	64	126	1,107	11.38	11.38
Ryan Stanton	VAN	D	54	95	773	12.29	11.32

PLAYER	TEAM	POS	GP	ESBK	USATA	EBS%	REBS%
Karl Alzner	WSH	D	82	127	1,134	11.20	11.20
Nate Guenin	COL	D	76	128	1,145	11.18	11.18
Alec Martinez	LAK	D	56	90	744	12.10	11.06
Deryk Engelland	CGY	D	76	121	1,097	11.03	11.03
Carl Gunnarsson	STL	D	61	96	822	11.68	11.03
Nate Prosser	MIN	D	63	80	628	12.74	10.99
Josh Gorges	BUF	D	46	101	897	11.26	10.93
Tom Gilbert	MTL	D	72	123	1,134	10.85	10.85
Michael Stone	ARI	D	81	149	1,378	10.81	10.81
Brooks Orpik	WSH	D	78	142	1,319	10.77	10.77
Zbynek Michalek	ARI/STL	D	68	104	970	10.72	10.64
Calvin de Haan	NYI	D	65	100	939	10.65	10.49
Willie Mitchell	FLA	D	66	105	1,010	10.40	10.40
David Schlemko	ARI/DAL/ CGY	D	44	69	568	12.15	10.37
Nick Schultz	PHI	D	80	118	1,141	10.34	10.34

[* Raw data obtained from War on Ice (blog), http://www.waronice.com.]

You surely have noticed that every player listed in these tables is a defenceman. This shouldn't surprise you—clearly shot-blocking is much more a blueliner's responsibility, and their positioning on the ice makes blocking shots easier as a result. The average NHL defenceman blocks about 8% of the opposing team's even-strength shots, while the average forward blocks only 3%. The variance is greater among forwards as well. While even the worst shot-blocking defenceman will still block about 4% of even-strength shots, some forwards block less than 1% of shots. These are offensive-minded wingers, such as Phil Kessel, Jiri Hudler, and Jeff Skinner, who will press the advantage in the offensive zone but are often still in the offensive zone when the opponents are finalizing their counterattack.

The very best shot-blocking forwards do so at a rate similar to an average defenceman, and while there are certainly some defensive specialists among the leaders, it's clear that blocking shots is not only the province of defensive-minded players.

NHL FORWARD LEADERS IN REBS%, 2013–14*

PLAYER	TEAM	POS	GP	ESBK	USATA	EBS%	REBS%
Boyd Gordon	EDM	C	74	67	836	8.01	7.21
Ben Smith	CHI	RW	75	48	669	7.17	5.84
Ryan Getzlaf	ANA	C	77	62	1,070	5.79	5.79
Mike Fisher	NSH	C	75	62	1,141	5.43	5.43
Nick Bonino	ANA	C	77	50	862	5.80	5.43
Tanner Glass	PIT	LW	67	43	657	6.54	5.38
Marcus Kruger	CHI	C	81	45	723	6.22	5.37
Michael Bournival	MTL	LW	60	40	587	6.81	5.30
Logan Couture	SJS	C	65	49	875	5.60	5.29
Ryan Callahan	NYR/TBL	RW	65	51	948	5.38	5.26

[* Raw data obtained from War on Ice (blog), http://www.waronice.com.]

NHL FORWARD LEADERS IN REBS%, 2014–15*

PLAYER	TEAM	POS	GP	ESBK	USATA	EBS%	REBS%
Ryan Getzlaf	ANA	C	77	79	975	8.10	7.98
Frans Nielsen	NYI	C	78	64	816	7.84	7.02
Boyd Gordon	EDM	C	68	57	656	8.69	6.85
Nick Bonino	VAN	C	75	63	854	7.38	6.79
David Backes	STL	RW	80	67	1,019	6.58	6.58
Joe Pavelski	SJS	C	82	64	983	6.51	6.46
Manny Malhotra	MTL	C	58	45	527	8.54	6.08
Matt Hendricks	EDM	LW	71	51	727	7.02	6.01
Lance Bouma	CGY	LW	78	59	1,010	5.84	5.84
Ben Smith	CHI/SJS	RW	80	50	756	6.61	5.82

[* Raw data obtained from War on Ice (blog), http://www.waronice.com.]

So with this information at hand, we can prepare a list of the very best shot-blockers in the NHL. This will not be a strict ranking—we won't list them one, two, three, since there is enough uncertainty involved here that making such precise determinations is foolhardy. Instead, we will provide a short list of the very best and some honourable mentions as well.

The Best Shot-Blocking Forwards

NICK BONINO: Bonino is a smart and solid two-way forward, and it should come as no surprise that he would be among the best in this category. He looks like a goalie when going down to block a shot, adopting something of a butterfly stance.

RYAN GETZLAF: Perhaps there's nothing that Getzlaf can't do on the ice. He's impressive offensively, of course, but he does not shirk his defensive duties.

He's smart and tough and clearly willing to sacrifice his body to prevent a scoring chance.

BOYD GORDON: Gordon came out of junior hockey as a solid two-way player, but in the professional ranks he has played a defensive role only, with a career high of eight goals in an NHL season. He's been a top penalty-killer, and if any forward deserves the title of best shot-blocker, it's probably him. He fearlessly throws his shins and knees at incoming pucks.

HONOURABLE MENTIONS: David Backes, Patrice Bergeron, Lance Bouma, Mike Fisher, Marcus Kruger, Manny Malhotra, Frans Nielsen, Joe Pavelski, and Ben Smith.

Even the best shot-blocking forwards can't compete with the blueliners, of course. Boyd Gordon may (or may not) be the best forward at blocking shots, but there's really no way he can be considered the best shot-blocker in the NHL. That title has to go to a defenceman, just by the nature of the job. If a forward was in position to block as many shots as a defenceman does, he really wouldn't be a forward, he'd be a defenceman.

The Best Shot-Blocking Defencemen

JOSH GORGES: Gorges uses his noggin to block shots, and not just in the literal sense, as in the famous clip of him against Washington in 2010, when the back of his helmet prevented a shot from reaching the net. He also uses his head in the sense of studying the angles, putting himself in the optimum position to block an opposing shot. And since the incident in 2010, he stays on his feet, which presumably makes his shot-blocking much less of a do-or-die play, since throwing your body to the ice can take you out of the play if it doesn't go as you expect.

ZBYNEK MICHALEK: Michalek was undrafted through his major junior career, but since that time he has established himself as the type of defenceman that every team should want. He's not super-physical, but he also doesn't take stupid penalties, and his competitiveness and hockey sense lead to an awful lot of blocked shots.

KRIS RUSSELL: Since arriving in Calgary in 2013, Russell has been a shot-blocking machine. Like every other defenceman of his size (5-foot-10), Russell had trouble earning respect for his defensive abilities, but when he was given first-pairing responsibilities he responded by blocking shots left, right, and centre. He uses his hockey smarts to lure opposing shooters into thinking they have an open lane and then takes it away from them. It's hard to rate anyone ahead of him given his results.

CHRISTOPHER TANEV: Sometimes it seems Tanev considers himself a second goalie, making kick saves and glove saves as a matter of course. He has an unorthodox style, but he's darn good at it.

MIKE WEBER: Weber is a classic stay-at-home defenceman. His play doesn't show up on the highlight reels, but he does what his team needs him to do in his own zone, and he isn't afraid to put himself in the line of 95-mile-per-hour slapshots. You have to admire that kind of disregard for personal health and safety, in some sense.

HONOURABLE MENTIONS: Johnny Boychuk, Ian Cole, Calvin de Haan, Raphael Diaz, Dan Girardi, Nate Guenin, Carl Gunnarsson, Barret Jackman, Alec Martinez, Willie Mitchell, Brooks Orpik, Nate Prosser, Nick Schultz, Ladislav Smid, Ryan Stanton, and Michael Stone.

Closing Thoughts

Blocked shots are rarely considered by hockey analysts due to the bias inherent in the raw numbers. But if you put them into the proper context, you can remove a lot of this bias and come up with a meaningful analysis of players. And while we did not produce a definitive ranking of the best shot-blockers in the NHL here, there are some candidates that certainly stand out from the others. Among the forwards, Boyd Gordon is probably the most prolific shot-blocker, while on defence it seems Kris Russell stands above the competition. It would be difficult to say you're wrong if you put these names forward as the elite shot-blockers in the NHL.

One thing we did not consider here is recording bias, the phenomenon whereby the official scorers in different NHL cities tend to have different definitions of what constitutes a certain statistic, such as a blocked shot or a hit. Press on to the next chapter, where we tackle this issue while looking at who the best NHL hitters are.

WHO IS THE BEST HITTER?

by IAIN FYFFE

Continuing our examination of certain much-maligned NHL stats, we're going to take a look at the best hitters in the league. The hits stat is a popular one in certain sections of the hockey media, but it doesn't get much attention from analysts or insiders. Arik Parnass took a stab at improving the usefulness of such stats, but such attempts tend to be few and far between.* So let's have a look at the NHL's heavy hitters to draw what information we can from this largely ignored statistic.

[* Arik Parnass, "Improving the NHL's Real Time Stats," *Habs Eyes on the Prize* (blog), October 31, 2013, http://www.habseyesontheprize.com/2013/10/31/5047828/improving-the-nhls-real-time-stats.]

NHL HIT LEADERS, 2012–13 TO 2014–15*

PLAYER	TEAM	GP	HITS
Matt Martin	NYI	205	968
Cal Clutterbuck	MIN/NYI	191	746
Ryan Reaves	STL	187	622
Derek MacKenzie	CBJ/FLA	196	617
Milan Lucic	BOS	207	607

PLAYER	TEAM	GP	HITS
David Backes	STL	202	598
Dustin Brown	LAK	207	591
Brooks Orpik	PIT/WSH	196	587
Chris Neil	OTT	162	587
Zac Rinaldo	PHI	157	586
Luke Schenn	PHI	184	575
Tommy Wingels	SJS	194	573
Steve Ott	BUF/STL	208	570
Tanner Glass	PIT/NYR	181	569
Alex Ovechkin	WSH	207	558
Kyle Clifford	LAK	199	540
Cody McLeod	COL	201	514
Cody Franson	TOR/NSH	202	511
Ryan Callahan	NYR/TBL	187	500
Brayden Schenn	PHI	211	497

[* Data obtained from the NHL's official website, htt p://www.nhl.com.]

The first step, as it should be when considering any counting stat, is to convert it into a rate. To improve the usefulness of the hits stat, we need some measure of opportunity in order to arrive at a relevant rate stat. With blocked shots, there is an obvious measure of opportunity, namely the number of shots directed toward the net when the player is on the ice. We have no equivalent with hits—there is no stat that measures how many times a player gets within range of a defending player while carrying the puck—so instead we need to make adjustments for opportunity as best we can.

The first such adjustment is fairly obvious, and it's based on the player's ice time. Clearly a player who's on the ice more often will get more opportunities to deliver hits than a player with less ice time. So in the end we will be presenting a player's adjusted hits on a per-minute basis in order to reduce the bias from playing time. Also, just as we did with blocked shots, we will be considering only even-strength events and playing time to eliminate any bias that might result from differing conditions in odd-man play.

The second adjustment is just as important as the first. Being on the ice isn't the only thing you need to have an opportunity to throw some hits; you also need the other team to have the puck. As we discussed in the chapter about blocked shots, this is the biggest reason these stats are largely ignored by those in the know: in order to get hits, the other team must have possession of the puck, and the other team having possession of the puck is generally a bad thing. But again, in a situation where the opponents already have the puck, knocking it loose with a bodycheck should be a valuable play. This adjustment is extremely important when dealing with hits, because there are a number of more-physical players in the NHL who throw lots of hits but are not good possession players, which inflates their hit totals relative to players who can control the puck.

A simple adjustment will get us where we want to go. We multiply each player's hits by his USAT%, which is his team's share of unblocked shot attempts when he's on the ice (a proxy for puck possession will be explained in greater detail in "Everything You

Ever Wanted to Know About Shot-Based Metrics [But Were Afraid to Ask]") and then divide by 50%. This provides the equivalent hits as if the player's team had 50% possession when the player was on the ice. Take hit monster Matt Martin's 2013–14 season for example. He recorded 356 even-strength hits, and he had a USAT% of 44.6, which suggests that the opponents had the puck about 55.4% of the time when he was on the ice. So his 356 hits are revised to 318 with this adjustment (356 times 0.446 and then divided by 0.5).

We'll also do a regression adjustment for players with less than 800 minutes, similar to what we did with blocked shots. And, finally, there is one more thing that absolutely needs to be considered here, in order to maximize the informativeness of the adjusted hits statistic. Hits, like any other NHL stat, are compiled by official scorers in each NHL city, and since there is no firm, absolute definition of what is and is not a hit, this situation leads to the possibility of recording bias. That is, some official scorers may award more hits than the others because they have a somewhat looser definition for what qualifies as a hit.

One simple way to approach this is to look at the hits that a team records on the road and compare them to the hits recorded at home. With the road figures being compiled in 29 different arenas with 29 (or more) different official scorers, the probability of recording bias creeping into the numbers is minimized. So road numbers can be seen as being more "true" than home numbers, and as such they can be used to estimate the bias inherent in the home numbers. Only one small adjustment is needed. Road teams have a possession disadvantage, with an average of 48.5% of shot events versus the home team's 51.5%. So in comparing road hits to home hits, we need to realize that a team should be expected to have fewer hits at home than on the road, all else being equal. Specifically, their home hits should be equal to about 94% of their road hits (48.5 divided by 51.5) if home and road hits were all being recorded on the same basis. So we can calculate the expected home hits by taking 94% of the road hits, and then we can calculate a recording bias by dividing the actual home hits to the expected home hits. This data is summarized in the following table. We use a three-year weighted average to calculate the bias adjustment, to minimize the chance of fluky results from a single season affecting things.

RECORDING BIAS FOR HITS, 2012–13 TO 2014–15*

TEAM	2015 HITS (HM-RD)	2014 HITS (HM-RD)	2013 HITS (HM-RD)	2015 BIAS	2014 BIAS	2013 BIAS	WT AVG BIAS	BIAS FACTOR
Anaheim	1,204-1,097	1,152-872	618-531	1.165	1.403	1.236	1.274	**1.137**
Arizona	1,459-1,030	1,213-960	727-533	1.504	1.342	1.448	1.428	**1.214**
Boston	1,008-1,104	1,072-936	668-532	0.970	1.216	1.333	1.141	**1.070**
Buffalo	994-1,210	953-1,039	572-507	0.872	0.974	1.198	0.978	**0.989**
Calgary	817-965	815-893	381-554	0.899	0.969	0.730	0.893	**0.947**
Carolina	1,139-814	1,004-803	602-498	1.486	1.328	1.284	1.382	**1.191**
Chicago	737-620	816-559	533-307	1.262	1.550	1.844	1.494	**1.247**
Colorado	979-1,039	1,018-1,017	546-695	1.001	1.063	0.834	0.992	**0.996**
Columbus	1,328-1,182	1,374-1,235	550-638	1.193	1.181	0.915	1.133	**1.066**
Dallas	897-760	882-762	670-529	1.253	1.229	1.345	1.262	**1.130**
Detroit	776-793	867-754	499-366	1.039	1.221	1.448	1.194	**1.097**
Edmonton	1,023-957	833-918	579-481	1.135	0.964	1.278	1.095	**1.048**
Florida	1,299-969	1,073-787	637-473	1.423	1.448	1.430	1.434	**1.217**
Los Angeles	1,368-1,262	1,487-1,122	837-609	1.151	1.407	1.459	1.315	**1.158**
Minnesota	552-771	642-770	477-479	0.760	0.885	1.057	0.870	**0.935**

TEAM	2015 HITS (HM-RD)	2014 HITS (HM-RD)	2013 HITS (HM-RD)	2015 BIAS	2014 BIAS	2013 BIAS	WT AVG BIAS	BIAS FACTOR
Montreal	877-819	868-854	511-556	1.137	1.079	0.976	1.082	**1.041**
Nashville	752-939	757-854	460-535	0.850	0.941	0.913	0.899	**0.950**
New Jersey	733-913	666-902	405-487	0.853	0.784	0.883	0.831	**0.916**
NY Islanders	1,437-1,248	1,025-1,067	519-499	1.223	1.020	1.104	1.118	**1.059**
NY Rangers	939-974	985-920	742-671	1.024	1.137	1.174	1.099	**1.050**
Ottawa	1,217-1,062	1,393-958	721-586	1.217	1.544	1.306	1.366	**1.183**
Philadelphia	1,409-1,102	1,162-1,012	670-646	1.358	1.219	1.101	1.251	**1.126**
Pittsburgh	1,371-1,124	1,120-1,042	587-642	1.295	1.141	0.971	1.169	**1.084**
San Jose	1,108-822	901-731	525-506	1.431	1.309	1.102	1.316	**1.158**
St. Louis	866-981	862-997	511-609	0.937	0.918	0.891	0.920	**0.960**
Tampa Bay	877-882	828-814	556-549	1.056	1.080	1.075	1.069	**1.035**
Toronto	1,423-1,122	1,544-1,048	905-721	1.347	1.564	1.333	1.431	**1.216**
Vancouver	691-891	580-911	508-512	0.824	0.676	1.054	0.811	**0.905**
Washington	1,160-1,175	1,027-877	586-505	1.048	1.243	1.232	1.163	**1.082**
Winnipeg	1,223-1,186	1,185-1,128	770-641	1.095	1.116	1.276	1.139	**1.070**
AVERAGE	**1,055-994**	**1,003-918**	**596-547**	**1.127**	**1.165**	**1.174**	**1.152**	**1.076**

[* Data obtained from the NHL's official website, accessed May 12, 2015, http://www.nhl.com.]

"Wt avg bias" is the weighted-average bias of the three seasons (2012–13, 2013–14, and 2014–15), assigning a weight of 50% to the first season and

100% to the latter two. Just by looking at the table you can see how persistent this particular bias tends to be. Teams with a positive bias (in excess of one) in one season typically have a positive bias in all three seasons (and mostly of the same magnitude), and teams with a negative bias (less than one) in one season typically have a negative bias in all three season (and mostly of the same magnitude). So it's clear that there are some significant biases here that need to be accounted for.

"Bias factor" is the multiplier you apply to the player's hit total for the season, calculated as the average of one and the weighted-average recording bias. This recognizes the fact that each player plays only half of his games at home, so half of his games should be recording-bias neutral. So a player with Chicago probably has a hits total that is 24.7% higher than it would be if all of his games were played using a neutral definition of hits, while a player with New Jersey probably has a hits total that is 8.4% lower than it should be.

Note that there is more than just simple recording bias to consider here. Not only do some official scorers award relatively more hits to all players, it seems clear that NHL official scorers generally award more hits to the home players than to the visiting players. You would expect that the home team would have about a 48:52 disadvantage in hits due to its possession advantage, but instead the average NHL home team has about a 52:48 advantage in hits. So not only is there recording bias, there is homerism involved as well, where the colour of a player's jersey can apparently influence whether or not a hit is awarded in a particular circumstance. Fortunately we can deal with both of these biases with a single adjustment, and this explains why the average bias factor is in excess of one. If there were only simple recording bias to deal with, you would expect the average bias adjustment to be very close to one, since some teams will be high and others will be low, offsetting each other. With hits, most teams are high, and some by a substantial amount.

We did not consider recording bias when discussing who the best shot-blocker is in the last chapter. While recording bias is always a potential issue due to the fact that humans do the official scoring, the chance of there being a significant bias with blocked shots is much lower than with hits. First of all, there is no apparent homerism in recording blocked shots. NHL home teams as a whole record about 96% of the blocked shots that road teams do, and this is to be expected given the possession advantage that home teams have. Secondly, the apparent bias that you can derive for blocked shots by using a method similar to the one just described is not of the same magnitude as it is for hits. Most teams are within 5% of what you would expect.

Finally, the most extreme teams in apparent recording bias for blocked shots tend to also be extreme teams, in the same direction, in recording bias for shot events. That is, teams that record a lower number of blocked shots at home than you would expect also tend to record a lower number of shots at goal than you would expect. If you take the six NHL teams that apparently have a positive bias of 5% or more in blocked shots (San Jose, Philadelphia, Montreal, Dallas, Toronto, and Washington) and compare them to shot-event results calculated by Michael Schuckers, you find that four of the six teams are also among the six teams with the highest positive shot-event recording bias. Similarly, of the seven teams with an apparent negative bias of 5% or more in recording blocked shots (New Jersey, Detroit, Columbus, Los Angeles, St. Louis, Boston, and Florida), five of them are among the six teams with the highest negative recording bias in shot events.*

[* Michael Schuckers and Brian Macdonald, "Accounting for Rink Effects in the National Hockey League's Real Time Scoring System," Statistical Sports Consulting, December 2, 2014, accessed May 12, 2015, http://arxiv.org/pdf/1412.1035v1.pdf.]

All of this is a long-winded way of saying that with respect to shot-blocking, the fact that a rate of blocked shots to shots directed at goal can be used eliminates a great deal of the recording bias that might otherwise be considered. Most of the apparent recording bias in blocked shots seems to arise from what is recorded as a shot, not from what is recorded as a block. We don't have anything similar to use for hits, however, and as such we need to consider recording bias.

We're now ready to look at the results of all of our adjustments and put together a list of the best hitters in the NHL from that data.

NHL LEADERS IN ADJUSTED EVEN-STRENGTH HITTING RATE (AESHT), 2013–14*

PLAYER	TEAM	POS	GP	ESHT	ESTOI	AESHT	AESHT PER 60
Matt Martin	NYI	LW	79	356	860	300	27.90
Cal Clutterbuck	NYI	RW	73	255	820	236	22.96
David Backes	STL	C	74	248	1,108	286	20.62
Chris Neil	OTT	RW	76	251	849	214	20.16
Colin McDonald	NYI	RW	70	216	822	192	18.71
Dustin Brown	LAK	RW	79	224	1,042	223	17.10
Derek MacKenzie	CBJ	C	71	233	661	227	17.02
Ryan Reaves	STL	RW	63	211	535	227	17.01
Boone Jenner	CBJ	C	72	208	946	196	16.58
Tuomo Ruutu	CAR/NJD	RW	76	212	971	198	16.32
Brandon Bollig	CHI	LW	82	202	820	166	16.16
Zac Rinaldo	PHI	C	67	229	481	215	16.12
Lance Bouma	CGY	C	78	178	808	162	16.02
Brandon Dubinsky	COL	C	76	218	1,065	210	15.74
Colin Greening	OTT	LW	76	210	920	180	15.67
Milan Lucic	BOS	LW	80	231	1,200	231	15.42
Nick Foligno	CBJ	LW	70	197	988	189	15.28
Blake Comeau	CBJ	LW	61	194	701	202	15.13
Radko Gudas	TBL	D	73	235	1,213	221	14.58
Ryan Callahan	NYR/TBL	RW	65	181	964	175	14.48

NHL LEADERS IN ADJUSTED EVEN-STRENGTH HITTING RATE (AESHT), 2014–15*

PLAYER	TEAM	POS	GP	ESHT	ESTOI	AESHT	AESHT PER 60
Matt Martin	NYI	LW	78	380	864	367	34.00
Cal Clutterbuck	NYI	RW	76	334	846	321	30.39
Derek MacKenzie	FLA	C	82	264	809	212	20.92

PLAYER	TEAM	POS	GP	ESHT	ESTOI	AESHT	AESHT PER 60
Ryan Reaves	STL	RW	81	276	686	272	20.39
Adam Lowry	WPG	LW	80	251	1,004	255	20.34
Lance Bouma	CGY	C	78	261	950	230	19.35
Tuomo Ruutu	NJD	LW	77	174	806	186	18.46
Kyle Clifford	LAK	LW	80	199	847	187	17.60
Blake Comeau	PIT	LW	61	178	856	174	16.23
Trevor Lewis	LAK	C	73	185	904	181	16.01
Tommy Wingels	SJS	C	75	241	1,027	202	15.70
Rob Klinkhammer	ARI/PIT/EDM	LW	69	236	765	209	15.66
Bryan Bickell	CHI	LW	80	202	907	175	15.47
Tom Wilson	WSH	RW	67	205	728	206	15.45
Nicolas Deslauriers	BUF	LW	82	255	910	175	15.41
Cody McLeod	COL	LW	82	260	770	205	15.36
Mark Borowiecki	OTT	D	63	210	861	165	15.32
Milan Lucic	BOS	LW	81	233	1,172	223	15.19
Zac Rinaldo	PHI	C	58	215	497	198	14.87
Michael Raffl	PHI	LW	67	161	859	158	14.72

[* Raw data obtained from War on Ice (blog), http://www.waronice.com.]

[* Raw data obtained from War on Ice (blog), http://www.waronice.com.]

You'll notice that these lists are dominated by forwards. This is a result of the nature of the positions. A defenceman does not have the freedom to pursue a puck carrier like a winger does; he has to be more conservative in order to avoid leaving his goaltender in the lurch. As such, it would be unfair to compare forwards and defencemen directly, so, like we did with blocked shots, we will look at the best from each group separately.

The Best Hitting Forwards

LANCE BOUMA: Bouma is the prototypical "energy" player; someone without a lot of hockey skill but who plays hard every shift and throws his body around with abandon. He blocks shots as well and does all this without taking too many penalties. The Flames may have overpaid him with his new contract, since the value he provides to a team is not difficult to replace, but not by too much.

CAL CLUTTERBUCK: Surely the owner of the best name in the NHL, especially for the type of player he is (only Rob Klinkhammer can really compete there), Clutterbuck hits often and hits hard. You might think that the presence of two Islanders so far above everyone else in terms of hits might suggest some bias remaining in the results, but Clutterbuck was already a big hitter with Minnesota before he went to New York.

DEREK MACKENZIE: MacKenzie's numbers give the appearance of a defensive specialist in lacrosse—very few offensive points but lots of hits and blocked shots and a very good faceoff man to boot.

MATT MARTIN: The undisputed king of hitting in the NHL. His high hit totals are not the result of being a poor possession player or having an official scorer biased in his favour. Removing as much bias as we can, he's still clearly out in front.

RYAN REAVES: Reaves is probably the "dirtiest" player in this group. It's a bit surprising that there are not more "dirty rat" players here,[11] but it does illustrate that you can play a very physical game at the NHL level without taking stupid penalties.

HONOURABLE MENTIONS: David Backes, Dustin Brown, Blake Comeau, Adam Lowry, Milan Lucic, Chris Neil, Zac Rinaldo, Tuomo Ruutu, and Tommy Wingels.

The Best Hitting Defencemen

ALEXEI EMELIN: Emelin is built like a brick and loves to hit, but he's a good example of why the value of a hit is questionable, at least among defencemen. In his pursuit of contact he often gets himself out of position, so sometimes he probably does more harm than good with a hit.

BROOKS ORPIK: A prototypical physical defenceman, Orpik doesn't do much with the puck but is good without it. He blocks shots and has enough speed to recover if one of his hits gets him out of position.

ROMAN POLAK: Polak is probably the biggest surprise in this group. He's a solid and steady defenceman but doesn't get that much recognition, even for his bodychecking.

---

11 Based on the definition of "dirty" penalties from Rob Vollman, Hockey Abstract 2014 (author, 2014), 273.

LUKE SCHENN: If you need to pick one blueliner as the best hitter, it would probably be Schenn. He has all-round defensive ability but hasn't really earned the complete trust of his coaches. He has averaged in excess of 22 minutes of ice time per game in a season, but he's also had three full seasons in which he has averaged less than 17 minutes per game. He's still fairly young, something that many people forget because he's been in the league since he was 18 years old, so there's still time for improvement.

HONOURABLE MENTIONS: Mark Borowiecki, Dustin Byfuglien, Jared Cowen, Simon Despres, Radko Gudas, Ben Lovejoy, Robyn Regehr, and Mark Stuart.

Closing Thoughts

Statistics like hits and blocked shots will probably never be useful enough to justify the aplomb with which many in the media celebrate them, but they also don't deserve to be summarily shunted to the dustbin by analysts. With the proper adjustments made and biases considered, you can massage some information out of them. Just don't overestimate their value.

WHO IS THE BEST PUCK STOPPER?

by ROB VOLLMAN

A goalie's most important job is to stop the puck. Overall there's a lot more to the position than that, like playing the puck, winning the shootout, overcoming fatigue and injury, playing consistently, drawing penalties, and raising one's game in the clutch, just to name a few. But if a goalie can't stop the puck, then none of those other talents really matters.

This chapter is not about who is the best goalie overall but is focused on which goalie is the best at stopping the puck—although they do tend to be one and the same. That's why the title of this chapter is meant as a precise descriptor of the upcoming analysis and not an effort to be hip—an attempt that would have failed on the first mention of regression or confidence intervals.

On the surface, finding the league's best puck stopper should be as easy as looking at the following leaderboard of each goalie's save percentage (SV%) over the past three years combined. After all, there's Tuukka Rask and Carey Price right at the top, with

Henrik Lundqvist not far behind. That sounds about right.

SAVE PERCENTAGE (SV%), 2012–13 TO 2014–15*

GOALIE	TEAM	SHOTS	SAVES	SV%
Cam Talbot	NYR	1,598	1,488	0.9312
Tuukka Rask	BOS	4,632	4,291	0.9264
Carey Price	MTL	4,799	4,438	0.9248
Cory Schneider	VAN/NJD	3,924	3,627	0.9243
Sergei Bobrovsky	CBJ	4,415	4,076	0.9232
Henrik Lundqvist	NYR	4,329	3,994	0.9226
Corey Crawford	CHI	3,963	3,652	0.9215
Steve Mason	CBJ/PHI	3,802	3,502	0.9211
Craig Anderson	OTT	3,491	3,214	0.9207
Ben Bishop	OTT/TBL	4,079	3,754	0.9203
Semyon Varlamov	COL	4,811	4,426	0.9200
Braden Holtby	WSH	4,642	4,269	0.9196
Roberto Luongo	VAN/FLA	3,883	3,565	0.9181
Jonathan Bernier	LAK/TOR	3,828	3,514	0.9180
Marc-André Fleury	PIT	4,486	4,116	0.9175
Frederik Andersen	ANA	2,219	2,036	0.9175
Devan Dubnyk	EDM/NSH/ARI/MIN	3,702	3,394	0.9168
Eddie Lack	VAN	2,253	2,065	0.9166
Alex Stalock	SJS	1,137	1,042	0.9164
Antti Niemi	SJS	4,736	4,339	0.9162
Michael Hutchinson	WPG	1,074	984	0.9162
Brian Elliott	STL	2,357	2,159	0.9160
Jaroslav Halak	STL/WSH/NYI	3,344	3,061	0.9154
Anton Khudobin	BOS/CAR	2,338	2,140	0.9153
Ryan Miller	BUF/STL/VAN	4,353	3,984	0.9152
Pekka Rinne	NSH	3,553	3,251	0.9150
Michal Neuvirth	WSH/BUF/NYI	1,951	1,785	0.9149
Robin Lehner	OTT	2,344	2,144	0.9147
Thomas Greiss	SJS/ARI/PIT	1,325	1,212	0.9147
Jonas Hiller	ANA/CGY	3,399	3,107	0.9141
Chad Johnson	BOS/NYI/BUF	1,304	1,192	0.9141
Jimmy Howard	DET	3,961	3,620	0.9139

GOALIE	TEAM	SHOTS	SAVES	SV%
James Reimer	TOR	3,090	2,824	0.9139
Jonathan Quick	LAK	3,968	3,625	0.9136
Kari Lehtonen	DAL	4,813	4,391	0.9123
Curtis McElhinney	CBJ	1,652	1,506	0.9116
Karri Ramo	CGY	1,943	1,771	0.9115
Jean-Sébastien Giguère	COL	1,077	981	0.9109
Jake Allen	STL	1,255	1,143	0.9108
Darcy Kuemper	MIN	1,539	1,401	0.9103
Mike Smith	ARI	4,764	4,334	0.9097
Ondrej Pavelec	WPG	4,248	3,858	0.9082
Al Montoya	WPG/FLA	1,358	1,233	0.9080
Jhonas Enroth	BUF/DAL	2,670	2,424	0.9079
Tim Thomas	FLA/DAL	1,390	1,262	0.9079
Carter Hutton	NSH	1,484	1,346	0.9070
Cam Ward	CAR	2,652	2,403	0.9061
Ben Scrivens	TOR/LAK/EDM	3,289	2,979	0.9057
Evgeni Nabokov	NYI/TBL	2,469	2,234	0.9048
Viktor Fasth	ANA/EDM	1,682	1,521	0.9043
Ray Emery	CHI/PHI	1,929	1,744	0.9041
Jonas Gustavsson	DET	1,043	943	0.9041
Anders Lindback	TBL/DAL/BUF	2,020	1,821	0.9015
Niklas Backstrom	MIN	2,070	1,866	0.9014
Justin Peters	CAR/WSH	1,426	1,285	0.9011
Reto Berra	CGY/COL	1,223	1,102	0.9011
Martin Brodeur	NJD/STL	1,794	1,616	0.9008
Ilya Bryzgalov	PHI/EDM/MIN/ANA	2,107	1,896	0.8999
Dan Ellis	CAR/DAL/FLA	1,290	1,157	0.8969
Jacob Markstrom	FLA/VAN	1,079	961	0.8906

Note: Only goaltenders who faced a minimum of 1,000 shots are included.

[* Raw goaltending data obtained from the NHL's website, accessed May 12, 2015, http://www.nhl.com.]

While this traditional view does take us at least halfway to the destination, there are a lot of factors to identify and unravel that can impact a goalie's save percentage to one extent or another, including random variation, manpower situation, shot quality, team effects, workload, and more.

Background

Goaltending has been extensively covered in the first two editions of this book, but there are reasons why it warrants a fresh look. First and foremost, it is a favourite topic among readers. Of everything we've written over the years, the goaltending analytics have garnered the most attention, have started the most discussions, have led to the most questions, and have motivated the most follow-up research.

In the inaugural edition of *Hockey Abstract,* an entire chapter was devoted to the hunt for the best overall goalie. Ultimately landing on Henrik Lundqvist, this chapter examined everything from catch-all statistics, Vezina voting, quality starts (which was further covered in a chapter of its own), even-strength save percentage, and differences between each starter and his backup. There was even a fun, whimsical look at how goalies perform when entering a game in relief.

In our second book, goaltending analytics grew to become a central theme. Two more chapters were devoted to the subject, one that examined what actually makes good goalies good and another that introduced and/or explored additional goaltending analytics, including home-plate save percentage, goal support, shootout save percentage, puck-handling, drawing penalties, clutch play, and the effect of workload. These analytics were used to answer questions like when goalies should be drafted, how much they should be paid, and when they should be pulled for an extra attacker. There was also a chapter devoted to shot quality, brief explorations of new models for goalie projections and league translations, and an examination of the impact of shot location and shooter talent on save percentage. There was even a silhouette of a goalie on the cover.

In this book, we construct goaltending age curves, and there is a study on replacement-level goaltending and a theory about how investing in goaltending is a crapshoot.

After all of this analysis, the demand for more has only continued to grow among our readers. Fellow analysts like Dawson Sprigings agreed, writing that "goaltending analysis is currently one of the most lacking subjects within hockey analytics."* In addition, TSN's Travis Yost argued that "if there's one area in hockey analytics that's lagging behind, it's primarily in the realm of goaltender studies,"** and Cam Lawrence added that "in the summer of analytics, goaltending has been noticeably absent from the discussions of major areas of advancement in our understanding of the sport we love."*** In short, there's still a nagging feeling of neglect throughout the community.

[* Dawson Sprigings (DTMAboutHeart), "Updated xSV%—Save Percentage Accounting for Shot Quality," *Don't Tell Me About Heart* (blog), June 3, 2015, http://donttellmeaboutheart.blogspot.ca/2015/06/updated-xsv-save-percentage-accounting.html.]

[** Travis Yost, "A New Way to Measure Goaltending Performance," TSN.ca, November 11, 2014, accessed May 12, 2015, http://www.tsn.ca/yost-a-new-way-to-measure-goaltending-performance-1.132438.]

[*** Cam Lawrence (Moneypuck), "Adjusted Save Percentage: Measuring the Impact Defense Has on Goaltending Statistics," *Canucks Army* (blog), October 16, 2014, http://canucksarmy.com/2014/10/16/adjusted-save-percentage-measuring-the-impact-defense-has-on-goaltending-statistics.]

That's why this chapter is going to be focused exclusively on save percentage, a topic that could probably fill an entire book all by itself. This choice isn't intended as an endorsement of the one-stat argument, which is a school of thought within the analytics community that all goaltending analysis should begin and end with this one number, but as an acknowledgement that it is the single best statistic we have for measuring a goalie's ability to stop the puck—once its applications and limitations are properly understood.

A great new aspect of this year's research is that not everything will have to be built from scratch. Years ago, there were so few people doing this kind of analysis publicly that all the data had to be collected manually and all-new statistics and perspectives had to be innovated. Now the field has grown to the point that almost everything that's required is already available, often in a form that's thoughtful and complete. We have definitely turned the corner in the world of hockey analytics.

The research has been organized into six broad groupings of the various factors that can affect save percentage. In decreasing order of importance, they include random variation, manpower situation, shot quality, team effects, workload, and then everything else. In each case, we're going to define the factor, break it down into smaller components, describe how it can be measured, establish its importance, and see how it changes our overall picture. Let's begin with the most important, and frustrating, of them all: puck luck.

Random Variation

NHL hockey is a sport of incredible skill, where only the top 0.06% of the world's hockey players compete in any given season. Despite the jaw-dropping level of talent involved, puck luck can have a tremendous impact on any given game and can continue to have a significant bearing on a goalie's save percentage even over an entire season.

To quickly demonstrate this reality, consider the chart on the next page, which shows the year-to-year consistency of each individual goalie's save percentage since 2005—or lack thereof. If luck were not a factor, then you would be able to set a Swiss watch by each goalie's consistent save percentage. Instead, you couldn't even set a sundial.

And that's where we begin. Statistically, there is very little relationship between a goaltender's save percentage from one season and the next, and that's mostly because of random variation. Why? Because shots are quite common but goals are relatively rare events, leaving even the most skilled goalie's numbers subject to a handful of bounces and breaks.

The importance of luck in save percentages is controversial because fans and analysts often feel they can actually explain such fluctuations after the fact. But if random variation tended to even out over an entire season, then a goalie's year-to-year save percentages would be closely related to one another, and most of the points in our chart would be hugging that line—or *any* line. Obviously NHL goalies have tremendous skill, but when random variation becomes a bigger part of the observed results than the player's underlying, persistent skill, then the points will be distributed in a broad, shotgun manner, exactly like on our chart.

YEAR-TO-YEAR GOALTENDING CONSISTENCY, 2005–06 TO 2014–15*

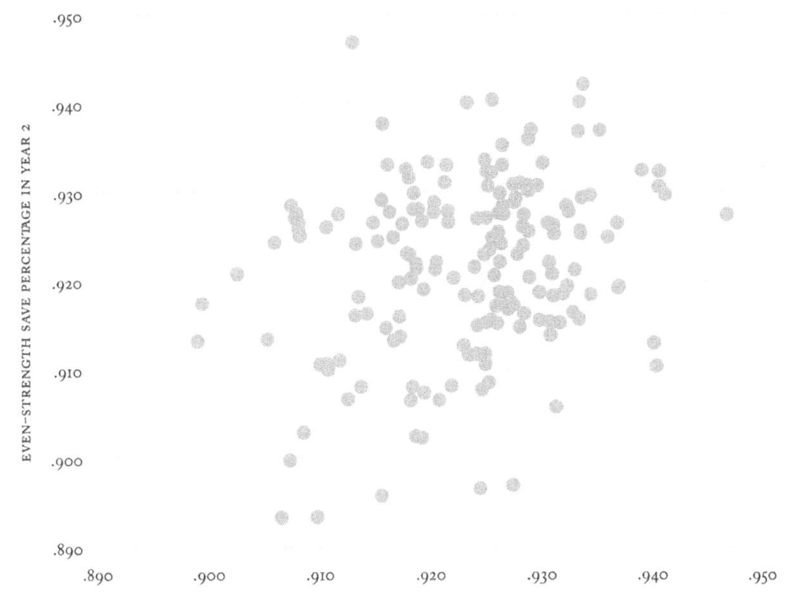

EVEN-STRENGTH SAVE PERCENTAGE IN YEAR 1

[* Raw goaltending data obtained from the NHL's website, accessed May 12, 2015, http://www.nhl.com.]

This is precisely why it's so hard to predict how well goalies and teams will do every year. Even the most consistent goaltender can experience random fluctuations resulting in at least a dozen unexpected goals, one way or the other. That can make or break both a goaltender's career and the fate of his team's entire season. Coaches have won the Jack Adams and others have been fired over nothing more than the fickle nature of puck luck.

To us, this random variation is also the single greatest obstacle in our efforts to use save percentage to identify the league's best puck stopper. After all, the central concept is to establish every goalie's theoretical "true talent" for stopping pucks, based on his observed performance over the years, and see who is best. However, each goalie's results are affected by random variation to an extent that exceeds the observed difference between them. In essence, we may ultimately land on the goalie who was luckiest (for lack of a better term), and not the one who is actually the best at stopping pucks.

Take Carey Price, for example. Even after over 10,000 career shots, we can say with only 95% confidence that his true performance level (at even strength) is somewhere between 0.923 and 0.933. That's a pretty broad range that puts Montreal's keeper anywhere from the league's best to merely average. That's hardly the kind of eye-opening discovery that justifies purchasing a book about hockey statistics, I know.

To demonstrate how dramatically each goalie's confidence intervals overlap, I was inspired by Brian Macdonald's recent work on the same subject,* and I constructed the following table of everyone's career save percentage at even strength (SV%) along with the lower and upper bounds of their true talent with 95% confidence.

[* Brian Macdonald, "A Bayesian Approach to Analyzing Goalies," *Greater Than Plus Minus* (blog), October 2013, http://www.greaterthanplusminus.com/p/goalies.html.]

CAREER EVEN-STRENGTH SAVE PERCENTAGE OF ACTIVE GOALIES*

GOALIE	TEAM	GP	SHOTS	LOW	HIGH	SV%
Tuukka Rask	BOS	266	6,191	0.9279	0.9403	0.9341
Braden Holtby	WSH	178	4,158	0.9215	0.9371	0.9293
Henrik Lundqvist	NYR	620	13,764	0.9244	0.9330	0.9287
Roberto Luongo	NYI/VAN/ FLA	864	19,490	0.9251	0.9323	0.9287
Carey Price	MTL	435	10,309	0.9230	0.9330	0.9280
Cory Schneider	VAN/NJD	212	4,647	0.9205	0.9353	0.9279
Sergei Bobrovsky	PHI/CBJ	230	5,425	0.9210	0.9348	0.9279
Pekka Rinne	NSH	381	8,934	0.9217	0.9325	0.9271
Jonas Hiller	ANA/CGY	378	8,490	0.9214	0.9325	0.9270
Corey Crawford	CHI	268	5,984	0.9186	0.9320	0.9253
Jaroslav Halak	MTL/STL/ WSH/NYI	331	7,390	0.9185	0.9305	0.9245
Jonathan Bernier	LAK/TOR	175	4,076	0.9161	0.9323	0.9242
Craig Anderson	CHI/FLA/ COL/OTT	406	9,620	0.9187	0.9293	0.9240

GOALIE	TEAM	GP	SHOTS	LOW	HIGH	SV%
Antti Niemi	CHI/SJS	338	7,734	0.9179	0.9298	0.9238
Jonathan Quick	LAK	407	8,531	0.9182	0.9294	0.9238
Semyon Varlamov	WSH/COL	267	6,595	0.9172	0.9300	0.9236
Kari Lehtonen	ATL/DAL	510	1,1927	0.9188	0.9283	0.9235
Ben Bishop	STL/OTT/ TBL	170	3,682	0.9148	0.9320	0.9234
Jimmy Howard	DET	338	7,569	0.9172	0.9292	0.9232
Ilya Bryzgalov	ANA/PHX/ PHI/EDM/ MIN	465	10,256	0.9176	0.9279	0.9228
James Reimer	TOR	175	4,329	0.9147	0.9306	0.9226
Anton Khudobin	MIN/BOS/ CAR	91	2,137	0.9105	0.9332	0.9219
Evgeni Nabokov	SJS/NYI/TBL	697	14,302	0.9175	0.9263	0.9219
Ryan Miller	BUF/STL/ VAN	604	14,188	0.9174	0.9263	0.9218
Niklas Backstrom	MIN	409	9,076	0.9158	0.9269	0.9213
Devan Dubnyk	EDM/NSH/ ARI/MIN	231	5,449	0.9134	0.9277	0.9205
Michal Neuvirth	WSH/BUF/ NYI	168	3,782	0.9118	0.9290	0.9204
Mike Smith	DAL/TBL/ ARI	387	8,941	0.9145	0.9258	0.9201
Martin Brodeur	NJD/STL	1,031	20,880	0.9164	0.9238	0.9201
Jhonas Enroth	BUF/DAL	131	3,168	0.9104	0.9293	0.9198
Steve Mason	CBJ/PHI	351	8,091	0.9137	0.9256	0.9197
Marc-André Fleury	PIT	595	1,2908	0.9142	0.9236	0.9189
Ray Emery	OTT/PHI/ ANA/CHI	287	5,802	0.9118	0.9258	0.9188
Robin Lehner	OTT	86	2,184	0.9070	0.9300	0.9185
Brian Elliott	OTT/COL/ STL	281	5,754	0.9109	0.9251	0.9180
Cam Ward	CAR	512	11,725	0.9124	0.9223	0.9174
Ondrej Pavelec	WPG	338	7,949	0.9106	0.9228	0.9167
Jason LaBarbera	NYR/LAK/ VAN/PHX/ EDM/ANA	187	3,800	0.9072	0.9249	0.9161
Anders Lindback	NSH/TBL/ DAL/BUF	111	2,426	0.9040	0.9262	0.9151

GOALIE	TEAM	GP	SHOTS	LOW	HIGH	SV%
Scott Clemmensen	TOR/FLA/NJD	191	4,006	0.9065	0.9238	0.9151
Dan Ellis	DAL/NSH/TBL/ANA/CAR/FLA	212	4,447	0.9068	0.9232	0.9150
Al Montoya	PHX/NYI/WPG/FLA	111	2,416	0.9023	0.9247	0.9135
Karri Ramo	TBL/CGY	122	2,757	0.9024	0.9235	0.9129
Ben Scrivens	TOR/LAK/EDM	129	3,027	0.9013	0.9216	0.9115
Curtis McElhinney	CGY/ANA/OTT/PHX/CBL	129	2,556	0.8985	0.9207	0.9096
Jonas Gustavsson	TOR/DET	148	3,223	0.8995	0.9193	0.9094

Note: Only goaltenders who faced a minimum of 2,000 shots are included.

[* Raw goaltending data obtained from the NHL's website, accessed May 12, 2015, http://www.nhl.com.]

I included only those who played in the NHL in 2014–15 and who have faced at least 2,000 career shots. Including sufficient data meant going all the way back to the 1997–98 season, when this data was first recorded by the NHL. However, the true talent level of some of the veteran goalies has changed a great deal since then, like Roberto Luongo, for example, during which time there's also been a slight change in scoring levels, so don't make any final evaluations based on this data.

The actual take-away point is just how close together all the goalies are. When you're dealing with the best 46 goalies in the entire world, you can never expect a great deal of separation from one individual to the next. Just as you'd struggle to make a distinction in the mushroom risottos prepared by the world's top chefs or differentiate between the world's best violinists, it can be virtually impossible for even an expert to isolate the subtle differences in the performances of the world's best goalies.

Even when using the largest data set possible, and arguably excessively so, there is still a great deal of overlap between goalies. Consider the top goaltender, Boston's Tuukka Rask. Is he really the best? Statistically we can't establish with 95% confidence that he's truly better than more than half the goalies on this list. Washington's Braden Holtby, in second place, has a lower bound that exceeds the upper bounds of only two goalies, journeyman backups Curtis McElhinney and Jonas Gustavsson.

This may not be particularly helpful information during Holtby's next contract negotiation, but it's important to understand how a lack of data and subsequent inability to overcome random variation is one of the main limitations of statistics. Instead of focusing on a black-and-white ordering of NHL goalies, random variation forces all questions to be reframed into matters of probabilities, like these:

- When the Stars signed free agent Antti Niemi, what were the chances they were getting someone superior to Kari Lehtonen?
- When Ottawa had to decide between keeping Craig Anderson, Ben Bishop, or Robin Lehner in the

summer of 2013, what were the odds that Anderson was the best?

- When Vancouver invested an extra $1.5 million in annual cap space to upgrade its goaltending from Roberto Luongo to Ryan Miller in the summer of 2014, what were the odds of an improvement?

In our case, the impact of random variation means that we cannot *truly* identify the league's top puck stopper but rather only who is most likely to be the league's best. While that may come as something of a disappointment, it does allow you to cling to the faint statistical hope that your favourite netminder is the league's best, virtually no matter who that may be. Unless it's McElhinney or Gustavsson, in which case you're right out of luck.

Manpower Situation

There's a misconception about why even-strength save percentage, instead of overall save percentage, is commonly used within the hockey analytics community. The problem isn't that there's something inherently useless about goaltending data in other manpower situations, but that power-play opportunities are so dangerous that even a short hot or cold streak can have a disproportionate impact on a goalie's year-end save percentage. Or, as Gabriel Desjardins once put it, "a significant portion of the variation in overall save percentage is variation in short-handed save percentage."*

[* Gabriel Desjardins, "Is It Possible to Determine Goaltender Talent on the PK?," *Arctic Ice Hockey* (blog), March 5, 2010, http://www.arcticicehockey.com/2010/3/5/1312360/is-it-possible-to-determine.]

In essence, the same problem exists with other types of highly dangerous shots, like breakaways for instance. The difference is that those types of shots are relatively rare, while about one in every six shots is taken on the power play, which has a far greater tendency to disrupt a goalie's overall numbers.

Take the interesting case of Tampa Bay's Ben Bishop and Nashville's Pekka Rinne, for example. These two goalies have virtually identical save percentages at even strength but are quite far apart overall. Why? Because of significant gaps in their short-handed save percentage and the percentage of shots they each faced in such situations—and these gaps exist even over three full seasons.

A TALE OF TWO GOALIES, 2012–13 TO 2014–15*

CATEGORY	BISHOP	RINNE
Overall SV%	0.920	0.915
Rank	10th	26th
Even-strength SV%	0.926	0.925
Rank	15th	16th
Short-handed SV%	0.900	0.855
Rank	6th	47th
Shots on PK	17.6%	13.2%
Rank	3rd	49th

[* Raw goaltending data obtained from the NHL's website, accessed May 12, 2015, http://www.nhl.com.]

At times, even-strength save percentage is admittedly chosen strictly out of habit, so it is important to identify when it should be used instead of overall save percentage. Earlier, when trying to quantify the impact of random variation, it made sense to eliminate whatever competing factors can be easily removed. When searching for the league's best puck stopper, however, it's important to use as much data as possible. Continuing this example, the value of having extra information about Bishop and Rinne outweighs the damage caused by the occasional and slight imbalance in penalty-killing time. In fact, there's even an argument to be made that we should somehow be including preseason and/or playoff data in our evaluations, too. Remember, Iain even used World Junior Championship data for his Projectinator.

Including penalty-killing data in an overall analysis is one thing, but evaluating goaltending talent using such numbers *by themselves* can be problematic. Even in Bishop's case, these are extremely small sample sizes, which are absolutely consumed by factors such as random variation and shot quality.

Consider the following chart, which compares each goalie's single-season save percentage at even strength (vertical axis) to his numbers on the penalty kill (horizontal axis). Not only is there the same big, shotgun mess, with no discernible pattern to establish any kind of relationship between the two variables, but there's also a much bigger spread horizontally than vertically.

COMPARING EVEN-STRENGTH TO PENALTY-KILLING SAVE PERCENTAGE, 2005–06 TO 2014–15*

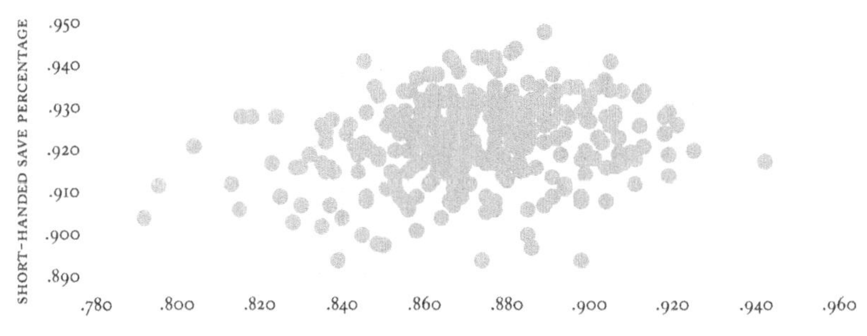

EVEN-STRENGTH SAVE PERCENTAGE

[* Raw goaltending data obtained from the NHL's website, accessed May 12, 2015, http://www.nhl.com.]

There really is no point in studying short-handed save percentage in isolation. This isn't to argue that no talent gap exists in short-handed save percentage, but rather that it is so small that it can't be captured from the limited sample of data. Even when trying to overcome that sample size issue by grouping all starters and backups together in *Hockey Abstract 2014,* Tom still observed that "there is no measurable difference between NHL goaltenders on the penalty kill."[12] The skill difference in NHL goalies is hard enough to capture at full strength, and we'd essentially have to compare Vezina winners to AHL backups before we'd statistically find a difference on the penalty kill.

Since we can't read the skill component in short-handed data, should we throw it out altogether? Absolutely not. As Tom put it, "penalty-killing save percentage is a skill, correlated to even-strength save percentage, but one that can hardly be observed."* Think of penalty-killing save percentage as the shy nerd at a high school dance. He is cool, he does have value, and just because he's not obvious doesn't mean he should be ignored completely. Well, that's my frame of reference at least.

---

12 Tom Awad, "What Makes Good Players Good, Part 3: Goaltenders," in Hockey Abstract 2014, by Rob Vollman (author, 2014), 79.

[* Tom Awad, "Numbers on Ice: How Much Skill Is There in Goaltending?," *Hockey Prospectus,* May 7, 2010, accessed May 12, 2015, http://www.hockeyprospectus.com/puck/article.php?articleid=558.]

CAUSES OF VARIANCE IN THREE-YEAR SAVE PERCENTAGE, 2007–08 TO 2009–10*

SCORE	SHOT QUALITY	LUCK	SKILL
Overall	12.6%	46.7%	40.7%
5-on-5 only	5.4%	58.0%	36.6%
4-on-5 only	19.9%	70.4%	9.7%

[* Tom Awad, "Numbers on Ice: How Much Skill Is There in Goaltending?," *Hockey Prospectus,* May 7, 2010, accessed May 12, 2015, http://www.hockeyprospectus.com/puck/article.php?articleid=558.]

Take a look at the following numbers. Though ravaged by the impact of shot quality and random variation (luck) when studied in isolation, including short-handed data in a goalie's overall save percentage does a better job of capturing a goalie's skill than using even-strength data by itself.

In the end, the picture won't really change that much whether an analysis uses short-handed data or not. Although the use of even-strength save percentage has practically become a tradition in the hockey analytics community, my feelings are that data is your friend and you should use as many data points as you possibly can, especially when you already have so few (data points, not friends). As long as there's an awareness of how it might affect the end results, there's nothing wrong with using overall save percentages.

Shot Quality

Should all shots be treated as being equally dangerous? One of the oldest and most passionate debates in the world of statistical hockey analysis is about shot quality (and how to pronounce Stanislav Neckar's name).

As summarized in Tom's dedicated chapter in *Hockey Abstract 2014,* the argument against shot quality has never been that it doesn't exist, that it averages out over a season, or that it isn't important, but rather that it is difficult to quantify, it is muddled by random variation and recording bias, and ignoring it usually won't lead anyone too far astray.[13]

If the impact of shot quality is important, why should it be ignored at all? The reality is that there isn't always time and space to cover every possible angle in a short TV/radio clip, magazine, or online article, especially for such complex concepts, so it makes sense to focus on only the simple and more important factors. The advantage of a book, on the other hand, is that the proper time and space can be invested to

---

13 Tom Awad, "Shot Quality," in Hockey Abstract 2014, by Rob Vollman (author, 2014), 32–43.

account for all factors—especially in studies like this, where that extra 5% accuracy can cause noticeable shifts in the end result.

So yes, shot quality most certainly exists, it is important, and in some situations it is more important than in others. As far back as 2008, universally acclaimed analyst Tore Purdy studied its year-to-year correlations and persistence, and he found that shot quality is "a *real* measurement of the average relative dangerousness of the shots allowed by any given team" and that it was "a reliable, enduring element of team defence."*

[* Tore Purdy (JLikens), "Shot Quality," *Objective NHL* (blog), October 26, 2008, http://objectivenhl.blogspot.ca/2008/10/shot-quality.html.]

While studying the impact of manpower situations on save percentage, we have seen how "most of the observed *skill* from stopping shots on the penalty kill comes from the quality of the shots," as Tom put it. In statistical terms, the importance of shot quality leaps from explaining 5.4% of the observed results at even strength to explaining 19.9% of the results in short-handed situations, compared to 36.6% and 9.7% for the goalie's own skill, respectively. In practical terms, that means that most of a goalie's observed success while killing penalties is actually the result of the team in front of him keeping shots to the outside and preventing screens and rebounds, rather than the result of his own talent at stopping pucks.

This greater role of shot quality with the man advantage also makes intuitive sense, given that such shots are typically taken by superior shooters, who take harder and more accurate shots, have more time and space to set up better attempts, and are consequently more likely to set up cross-ice passes, screens, deflections, and rebounds. It's a goalie's nightmare.

It's not just while killing penalties that observable changes in shot quality occur. It also impacts the goalie on the team with the man advantage, and it has been clearly observed to change in close-score situations in the final period, when the trailing team starts to really open things up and gamble more often.

Even in normal, full-strength situations, shot quality does not come out in the wash—it is the red sock that gets mixed in with the whites. It can vary from team to team and from goalie to goalie, and it has been known to sabotage otherwise sound theories, studies, and predictions.

While it can be a challenging factor to sort out (to say the least), the impact of shot quality on a goalie's save percentages makes a lot more sense when it is broken down into its components. Most prominently, that includes the location from which the shot was taken, its speed and accuracy, and whether or not it was generated off a high-danger situation, such as on the rush, after a cross-ice pass, or on a deflection or rebound. Let's examine each one separately and in more detail. Oh, and it's pronounced "stan-is-lav nets-cash."

Shot Location

The most important and popular component of shot quality is the shot's location. The two terms are sometimes used synonymously, given that shot location is one of the few components of a shot's quality that is actually recorded in NHL game files—although not always with the same accuracy and consistency as a Stamkos wrister.

All things being equal, it makes perfect sense that shots become more dangerous the closer they are taken to the net. After all, the closer someone takes a shot, the more accurately he can aim it and the less time the goalie will have to react to it. In fact, the difference in shooting percentage between forwards and defencemen, which is 9.3% and 4.1% respectively, can be almost completely explained by shot distance.

That's why heat maps, like the kind we included in *Hockey Abstract 2014,* show an almost linear relationship between shot distance and shooting percentage.[14] Shots taken at a distance of 26 feet have a roughly average 9% chance of scoring, and

---

14 Tom Awad, "Shot Quality," in Hockey Abstract 2014, by Rob Vollman (author, 2014), 34–43.

the odds go all the way up to around 20% as the distance approaches 10 feet or less and drop down to under 4% as it begins to exceed 40 feet.

Distance has an obvious impact on an individual shot-by-shot basis, but to what extent does it impact the numbers over a whole season? Recently, the average shot distance one individual goalie faces compared to another can easily vary by up to 4 feet, which still represents a swing of three or four points in his final save percentage.

Even in even-strength situations, where shot quality has been observed to have its smallest impact, "shot distance—and by extension, long-run shot quality allowed—accounts for 5% of save percentage," according to a study by Gabriel Desjardins.* Given the incredible parity throughout the league, that small impact is worth considering whenever save percentage is used to compare goalies.

[* Gabriel Desjardins, "Shot Distance Allowed as a Team Talent," *Arctic Ice Hockey* (blog), October 24, 2011, http://www.arcticicehockey.com/2011/10/24/2506209/shot-distance-allowed-as-a-team-talent.]

To that end, I previously introduced a new statistic called home-plate save percentage, which refers to a goalie's save percentage exclusively against shots taken inside the so-called home-plate area.[15] Also known as "the House," the home-plate area is the diamond shape created by drawing an imaginary line from each goalpost to the respective faceoff dots, then directly to the top of the circle, and across. Shots taken within this area are the game's most dangerous, and judging goalies in this fashion serves to equalize the playing field and more easily permits direct comparison.

Over the past three seasons combined, most individual goalies faced about 37.5% of their shots from inside this home-plate area, posting an average save percentage of about 0.83 on the shots that were taken from within and 0.96 on those from the outside. While a goalie's ability to stop shots from the outside isn't meant to be ignored, it's really his ability to stop the dangerous shots that indicates his true value to a team. For example, Devan Dubnyk's early success on inside shots served to foreshadow his breakout season in 2014–15.

In the following leaderboard, the shots, goals, and save percentage from the outside (out) are listed first, followed by how they performed inside the home-plate area (HP), separated by the percentage of shots each goalie faced from inside the home-plate area (HP%),

---

15 Rob Vollman, "Goaltending Analytics Revisited," in Hockey Abstract 2014 (author, 2014), 88.

which shows how some goalies have definitely had it rougher than others.

HOME-PLATE SAVE PERCENTAGE (HP SV%), 2012–13 TO 2014–15*

GOALIE	TEAM	OUT SHOTS	OUT GOALS	OUT SV%	HP%	HP SHOTS	HP GOALS	HP SV%
Cam Talbot	NYR	961	29	0.970	39.9%	637	81	0.873
Thomas Greiss	SJS/ ARI/ PIT	785	40	0.949	40.8%	540	73	0.865
Braden Holtby	WSH	2,896	127	0.956	37.6%	1,746	246	0.859
Cory Schneider	VAN/ NJD	2,368	76	0.968	39.7%	1,556	221	0.858
Henrik Lundqvist	NYR	2,618	88	0.966	39.5%	1,711	247	0.856
Tuukka Rask	BOS	3,051	111	0.964	34.1%	1,581	230	0.855
Jaroslav Halak	STL/ WSH/ NYI	1,930	75	0.961	42.3%	1,414	208	0.853
Devan Dubnyk	EDM/ NSH/ ARI/ MIN	2,253	91	0.960	39.1%	1,449	217	0.850
Sergei Bobrovsky	CBJ	2,784	92	0.967	36.9%	1,631	247	0.849
Semyon Varlamov	COL	2,940	101	0.966	38.9%	1,871	284	0.848
Ben Bishop	OTT/ TBL	2,468	80	0.968	39.5%	1,611	245	0.848
Carey Price	MTL	3,018	89	0.971	37.1%	1,781	272	0.847
Jake Allen	STL	755	33	0.956	39.8%	500	79	0.842
Brian Elliott	STL	1,473	58	0.961	37.5%	884	140	0.842
Eddie Lack	VAN	1,323	40	0.970	41.3%	930	148	0.841
Jonathan Bernier	LAK/ TOR	2,410	88	0.963	37.0%	1,418	226	0.841
Michael Hutchinson	WPG	642	21	0.967	40.2%	432	69	0.840

GOALIE	TEAM	OUT SHOTS	OUT GOALS	OUT SV%	HP%	HP SHOTS	HP GOALS	HP SV%
Kari Lehtonen	DAL	2,841	107	0.962	41.0%	1,972	315	0.840
Corey Crawford	CHI	2,458	70	0.972	38.0%	1,505	241	0.840
Evgeni Nabokov	NYI/ TBL	1,460	73	0.950	40.9%	1,009	162	0.839
Steve Mason	PHI	2,521	94	0.963	33.7%	1,281	206	0.839
Frederik Andersen	ANA	1,305	36	0.972	41.2%	914	147	0.839
Jonas Hiller	ANA/ CGY	2,092	81	0.961	38.5%	1,307	211	0.839
Jhonas Enroth	BUF/ DAL	1,687	85	0.950	36.8%	983	161	0.836
Roberto Luongo	VAN/ FLA	2,426	78	0.968	37.5%	1,457	240	0.835
Chad Johnson	ARI/ BOS/ NYI	765	23	0.970	41.3%	539	89	0.835
Michal Neuvirth	WSH/ BUF/ NYI	1,200	41	0.966	38.5%	751	125	0.834
Ryan Miller	BUF/ STL/ VAN	2,733	99	0.964	37.2%	1,620	270	0.833
Anton Khudobin	BOS/ CAR	1,421	45	0.968	39.2%	917	153	0.833
Jean-Sébastien Giguère	COL	673	28	0.958	37.5%	404	68	0.832
Craig Anderson	OTT	2,330	80	0.966	33.3%	1,161	197	0.830
Karri Ramo	CGY	1,165	39	0.967	40.0%	778	133	0.829
Jimmy Howard	DET	2,529	96	0.962	36.2%	1,432	245	0.829
Robin Lehner	OTT	1,544	63	0.959	34.1%	800	137	0.829
Mike Smith	ARI	2,930	115	0.961	38.5%	1,834	315	0.828
Marc-André Fleury	PIT	2,790	77	0.972	37.8%	1,696	293	0.827
Antti Niemi	SJS	3,041	101	0.967	35.8%	1,695	296	0.825
Alex Stalock	SJS	697	18	0.974	38.7%	440	77	0.825
Reto Berra	CGY/ COL	727	34	0.953	40.6%	496	87	0.825
James Reimer	TOR	2,040	81	0.960	34.0%	1,050	185	0.824
Al Montoya	WPG/ FLA	853	36	0.958	37.2%	505	89	0.824

GOALIE	TEAM	OUT SHOTS	OUT GOALS	OUT SV%	HP%	HP SHOTS	HP GOALS	HP SV%
Jonathan Quick	LAK	2,548	91	0.964	35.8%	1,420	252	0.823
Dan Ellis	CAR/ DAL/ FLA	758	38	0.950	41.2%	532	95	0.821
Martin Brodeur	NJD/ STL	1,125	57	0.949	37.3%	669	121	0.819
Ondrej Pavelec	WPG	2,629	97	0.963	38.1%	1,619	293	0.819
Viktor Fasth	ANA/ EDM	1,053	46	0.956	37.4%	629	115	0.817
Ben Scrivens	TOR/ LAK/ EDM	2,096	89	0.958	36.3%	1,193	221	0.815
Carter Hutton	NSH	929	34	0.963	37.4%	555	104	0.813
Cam Ward	CAR	1,703	69	0.959	35.8%	949	180	0.810
Tim Thomas	FLA/ DAL	879	31	0.965	36.8%	511	97	0.810
Pekka Rinne	NSH	2,391	81	0.966	32.7%	1,162	221	0.810
Ilya Bryzgalov	PHX/ EDM/ MIN/ ANA	1,352	65	0.952	35.8%	755	146	0.807
Curtis McElhinney	CBJ	1,062	31	0.971	35.7%	590	115	0.805
Anders Lindback	TBL/ DAL/ BUF	1,269	51	0.960	37.2%	751	148	0.803
Ray Emery	CHI/ PHI	1,224	46	0.962	36.5%	705	139	0.803
Darcy Kuemper	MIN	1,027	36	0.965	33.3%	512	102	0.801
Justin Peters	CAR/ WSH	860	28	0.967	39.7%	566	113	0.800
Jonas Gustavsson	DET	697	25	0.964	33.2%	346	75	0.783
Jacob Markstrom	FLA/ VAN	681	28	0.959	36.9%	398	90	0.774
Niklas Backstrom	MIN	1,422	56	0.961	31.3%	648	148	0.772

Note: Only goaltenders who faced a minimum of 1,000 shots are included.

[* Data obtained from Greg Sinclair, *Super Shot Search,* http://somekindofninja.com/nhl/(site discontinued).]

To put it in different terms, the following chart uses the same data, with each goalie's home-plate save percentage on the vertical axis and his save percentage from the outside on the horizontal axis. The goalies who are strongest up close will be at the top of the chart, while those best at handling outside shots are to the right.

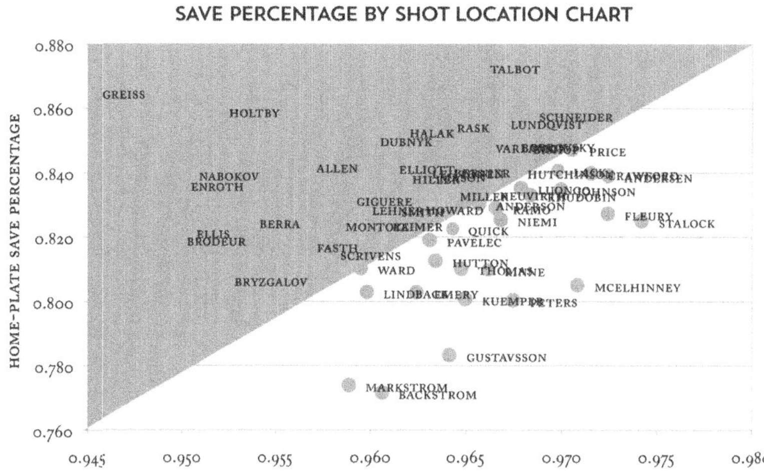

SAVE PERCENTAGE FROM OUTSIDE THE HOME-PLATE AREA

How does this change our current view of the league's best puck stopper?

- It further reinforces the case to include Washington's Braden Holtby and New Jersey's Cory Schneider in the picture, along with known elite goalies like Boston's Tuukka Rask and the Rangers' Henrik Lundqvist.
- It also elevates a few names, like the Islanders' Jaroslav Halak and Tampa Bay's Ben Bishop, into that second level with near-elite goalies like Columbus's Sergei Bobrovsky and Montreal's Carey Price (surprisingly).

- Among backups and platoon goalies, it showcases the potential of Edmonton's Cam Talbot and establishes the Islanders' Thomas Greiss as perhaps the league's next Devan Dubnyk.
- It all but destroys the case for Chicago's Corey Crawford, Philadelphia's Steve Mason, and especially Nashville's Pekka Rinne, who appear to rely on shots being forced to the outside.

This new perspective quickly inspired a flurry of new and refined developments, one of which is the adjusted save percentage developed by Sam Ventura and Andrew Thomas of *War on Ice,* who were subsequently hired by the Pittsburgh Penguins and Minnesota Wild, respectively.*

[* Sam Ventura, "Adjusted Save Percentage: Taking into Account High, Medium, and Low Probability Shots," *War on Ice* (blog), November 5, 2014, http://blog.war-on-ice.com/adjusted-save-percentage-takinginto-account-high-medium-and-low-probability-shots/.]

Initially moving forward with the same home-plate area, Ventura and Thomas found that there was actually a small area just outside it, directly in front of the net, where shooting percentages remained superior. They also wanted to separate out shots taken in the low slot, because they had double the shooting percentage as those taken throughout the rest of the expanded home-plate area. That left them with three areas instead of two, which they blended into their new adjusted save percentage using the same league-average ratio for every goalie.

Bear in mind that the following leaderboard uses their new nomenclature. Outside shots are termed low-danger shots, the expanded home-plate area is considered to be a medium-danger zone, except for the low slot, which is dubbed the high-danger zone.

ADJUSTED SAVE PERCENTAGE (ADJSV%), 2012–13 TO 2014–15*

GOALIE	TEAM	LOW	SV%	MED	SV%	HIGH	SV%	ADJSV%
Tuukka Rask	BOS	2,086	0.982	1,193	0.935	999	0.859	0.9412
Carey Price	MTL	2,064	0.979	1,102	0.934	941	0.851	0.9372
Henrik Lundqvist	NYR	2,277	0.981	1,112	0.925	1,089	0.854	0.9371
Jimmy Howard	DET	1,525	0.981	867	0.943	802	0.834	0.9369
Cory Schneider	VAN/NJD	1,302	0.976	796	0.939	707	0.849	0.9365
Cam Talbot	NYR	625	0.972	299	0.934	305	0.862	0.9365
Craig Anderson	OTT	1,428	0.977	716	0.947	639	0.836	0.9360
Chad Johnson	ARI/BOS/NYI	415	0.986	275	0.929	254	0.836	0.9360
Sergei Bobrovsky	CBJ	1,595	0.974	897	0.938	751	0.846	0.9345

GOALIE	TEAM	LOW	SV%	MED	SV%	HIGH	SV%	ADJSV%
Jonas Hiller	ANA/ CGY	1,323	0.973	746	0.935	744	0.851	0.9345
Jonathan Bernier	LAK/TOR	1,229	0.975	784	0.940	689	0.840	0.9345
Ben Bishop	OTT/TBL	1,478	0.976	917	0.936	819	0.841	0.9340
Steve Mason	CBJ/PHI	1,447	0.975	779	0.950	662	0.827	0.9336
Corey Crawford	CHI	2,066	0.975	1,107	0.926	1,067	0.849	0.9332
Braden Holtby	WSH	1,856	0.973	991	0.928	876	0.853	0.9332
Jonathan Quick	LAK	1,715	0.981	1,042	0.929	944	0.833	0.9328
Thomas Greiss	SJS/ARI/ PIT	440	0.961	254	0.924	239	0.882	0.9328
Semyon Varlamov	COL	1,633	0.973	970	0.933	987	0.845	0.9324
Marc-André Fleury	PIT	1,740	0.978	944	0.932	920	0.827	0.9309
Anton Khudobin	BOS/CAR	818	0.983	474	0.928	452	0.820	0.9308
Kari Lehtonen	DAL	1,533	0.971	992	0.928	938	0.845	0.9304
Eddie Lack	VAN	783	0.975	464	0.941	417	0.821	0.9301
Roberto Luongo	VAN/FLA	1,424	0.973	784	0.928	621	0.838	0.9296
Devan Dubnyk	EDM/ NSH/ARI/ MIN	1,286	0.968	806	0.926	719	0.850	0.9295
Frederik Andersen	ANA	1,032	0.979	564	0.937	487	0.813	0.9292
Jaroslav Halak	STL/ WSH/NYI	1,167	0.966	693	0.927	687	0.851	0.9290
Jhonas Enroth	BUF/DAL	915	0.968	541	0.943	498	0.829	0.9287
Mike Smith	ARI	1,536	0.972	883	0.912	918	0.852	0.9286
Karri Ramo	CGY	714	0.973	414	0.928	427	0.832	0.9285
James Reimer	TOR	1,154	0.970	668	0.918	654	0.849	0.9283
Michal Neuvirth	WSH/ BUF/NYI	657	0.976	364	0.906	321	0.845	0.9275
Ryan Miller	BUF/STL/ VAN	1,476	0.975	939	0.921	836	0.831	0.9273
Brian Elliott	STL	918	0.969	515	0.938	405	0.825	0.9273
Antti Niemi	SJS	1,897	0.977	1,010	0.928	862	0.817	0.9271
Pekka Rinne	NSH	1,473	0.977	753	0.922	582	0.822	0.9267
Ray Emery	CHI/PHI	671	0.971	361	0.921	376	0.832	0.9256
Evgeni Nabokov	NYI/TBL	775	0.971	555	0.920	553	0.829	0.9249
Jake Allen	STL	434	0.956	292	0.927	268	0.851	0.9238

GOALIE	TEAM	LOW	SV%	MED	SV%	HIGH	SV%	ADJSV%
Ondrej Pavelec	WPG	1,456	0.973	867	0.924	758	0.815	0.9236
Robin Lehner	OTT	812	0.963	441	0.909	400	0.842	0.9211
Darcy Kuemper	MIN	676	0.976	313	0.892	264	0.833	0.9209
Justin Peters	CAR/ WSH	467	0.981	285	0.931	268	0.779	0.9208
Ben Scrivens	TOR/ LAK/ EDM	1,123	0.966	702	0.930	571	0.813	0.9206
Ilya Bryzgalov	PHI/ EDM/ MIN/ANA	768	0.970	464	0.919	389	0.816	0.9205
Cam Ward	CAR	919	0.964	544	0.927	460	0.817	0.9200
Al Montoya	WPG/ FLA	471	0.963	261	0.906	260	0.839	0.9196
Anders Lindback	TBL/ DAL/BUF	693	0.969	462	0.926	361	0.797	0.9176
Tim Thomas	FLA/DAL	474	0.977	266	0.902	224	0.803	0.9172
Viktor Fasth	ANA/ EDM	587	0.964	293	0.891	337	0.840	0.9165
Curtis McElhinney	CBJ	619	0.975	277	0.902	256	0.803	0.9160
Carter Hutton	CHI/NSH	548	0.977	305	0.900	217	0.801	0.9159
Martin Brodeur	NJD/STL	616	0.957	346	0.908	291	0.834	0.9154
Dan Ellis	CAR/ DAL/FLA	406	0.958	256	0.934	242	0.801	0.9146
Niklas Backstrom	MIN	723	0.969	403	0.914	324	0.790	0.9130

Note: Only goaltenders who faced a minimum of 1,000 shots are included.

[* Data obtained from *War on Ice* (blog), http://www.waronice.com.]

Adjusting for shot location doesn't change the picture by much, but it *does* change the picture, and arguably closer toward conventional wisdom.

- Tuukka Rask and Henrik Lundqvist are back on top, and Carey Price is back in the picture between them, where most pundits would argue that he belongs.

- Cory Schneider remains in the picture, but Braden Holtby has tumbled back into the pack.
- Jonas Hiller, Corey Crawford, and Steve Mason are back at that second "Sergei Bobrovsky" level, to which Craig Anderson has been elevated.
- Among backups, which the lack of data makes especially volatile, Chad Johnson is elevated above Thomas Greiss and just below Cam Talbot.
- Other big-name goalies, like Roberto Luongo and Pekka Rinne, remain out of the picture.

That's a pretty accurate assessment, in my view. While additional precision requires far greater complexity, it is still quite possible to go further yet. Michael Schuckers, with his defence-independent goaltending rating (DIGR), and Wesley Yue, with his location-adjusted expected-goals percentage (LAEGP), went beyond dividing save percentage into two or three categories, and they actually assigned a weighting to every single spot on the ice.* The odds of each shot going in were calculated based on the historic success rate from that exact spot on the ice and those nearby.

[* Michael E. Schuckers, "DIGR: A Defense Independent Rating of NHL Goaltenders Using Spatially Smoothed Save Percentage Maps" (St. Lawrence University and Statistical Sports Consulting, March 4, 2011), accessed May 9, 2015, http://myslu.stlawu.edu/~msch/sports/Schuckers_DIGR_MIT_2011.pdf ;

Wesley Yue, "Introducing a New Stat: Location Adjusted Expected Goal Percentage," Hockey Metrics, August 28, 2013, Web, http://hockeymetrics.net/introducing-a-new-stat-location-adjusted-expectedgoals-percentage/.]

The two statistics differ slightly in how they use that information. DIGR calculates what each goalie's save percentage would be had he faced the same distribution of shots as the league average, while LAEGP is the number of goals each netminder prevented relative to what you'd expect based on the location of the shots he faced.

Neither statistic has results that are readily available for public consumption, and the results that have been made available don't stray too dramatically from what Ventura, Thomas, and I have found. However, they do demonstrate just how far NHL teams can go to get that extra level of precision.

Speed and Accuracy

There is much more to overall shot quality than the location from which the shot was taken. Players take shots of different speeds, and some snipers can pick the corners and holes a lot more reliably than others. Furthermore, some shots are taken off the rush or off a cross-ice pass and possibly with screens and/or involving deflections and rebounds. That's why Andrew Thomas of *War on Ice* is among those who have found that the increased shooting percentage on the power play cannot be fully explained by shot location. Instead, he discovered that shooting percentages in all locations increased noticeably and consistently with the man advantage, in part because of the improved speed and accuracy of the shots themselves.

SHOOTING PERCENTAGE BY SHOT DANGER AND MANPOWER SITUATION, 2008–09 TO 2013–14*

SCORE	LOW DANGER	MEDIUM DANGER	HIGH DANGER
Even strength	2.7%	7.0%	17.0%
Power play	6.7%	12.2%	21.9%

[* Andrew Thomas, "The Road to WAR, Part 10: Modern Goaltending and Shooting," *War on Ice* (blog), April 3, 2015, http://blog.war-on-ice.com/the-road-to-war-part-10-modern-goaltending-and-shooting/.]

We have a nice theory that explains these results, but we have no evidence that improved shot speed and accuracy can boost shooting percentages by 4% or 5%. Without access to any such information in NHL game files, we don't even know for a fact that power-play shots actually are harder or more accurate.

Lacking the specific data, can the speed and accuracy of a shot be inferred from the career shooting percentage of the shooter himself? For example, shots taken by Steven Stamkos are probably faster and more accurate than those taken by Shawn Thornton, but to what extent? The same factors that affect save percentages apply to shooting percentages as well, meaning that most of the difference between their career numbers of 17.2% and 4.4% could actually be the result of random variation, manpower situation, and shot location.

Thankfully, we have Matt Pfeffer to help sort some of it out. To predict the scoring chance of any particular shot, Pfeffer built a model that integrated as many predictive elements as he could extract from NHL game files, such as the location the shot was taken, the time that had elapsed since the last shot, the type of shot, the manpower situation, the current score, the game time elapsed, and so on.*

[* Matt Pfeffer, "Exploring the Theoretical Limits of Shot Quality," *Hockey Prospectus* (blog), June 4, 2014, http://www.hockeyprospectus.com/exploring-the-theoretical-limits-of-shot-quality/.]

In theory, the speed and accuracy of a player's shot can be calculated as the difference between his actual shooting percentage and what's expected from Pfeffer's model, since those are the two most significant factors remaining. The results? "The best shooters (Tanguay, Marchand, Steven Stamkos) can add 3–5% in shooting percentage above what the model inputs predict." That is to say, a hypothetical shot that would normally be expected to go in 8% of the time, given its location and other factors, would go in 11% to 13% of the time if it is taken by an elite shooter like Stamkos.*

[* Matt Pfeffer, "Exploring the Theoretical Limits of Shot Quality," *Hockey Prospectus* (blog), June 4, 2014, http://www.hockeyprospectus.com/exploring-the-theoretical-limits-of-shot-quality/.]

This result comes stunningly close to explaining the *entire* gap Thomas found between even-strength and power-play shooting percentages after accounting for shot location. It could just be a coincidence that the two results match up, but it's more likely to be further evidence that the increase in power-play shooting percentages is a result of better shooters taking harder and more accurate shots.

This theory makes a lot of intuitive sense on an individual basis, but is there really enough difference in the average speed and accuracy of shots across the entire league to significantly affect a goalie's save percentage? After all, there's only one Steven Stamkos, and it isn't as if the same goalie has to face Tampa Bay night after night. In the age of cap-imposed parity, it stands to reason that the average speed and accuracy of opposing shooters should mostly average out over the course of a full season.

In our second book, I tackled this larger question by calculating each goalie's save percentage relative to the average shooting percentage of the opposing teams he faced.[16] The premise was that those who tended to face teams with higher shooting percentages

---

16 Rob Vollman, "Goaltending Analytics, Revisited," in Hockey Abstract 2014 (author, 2014), 92.

were likely facing shots of greater speed and accuracy, which would affect their save percentages.

This proved to be a little too tricky, however, because each team's shooting percentage is impacted by so many of the same factors that affect save percentages, most notably random variation, which is responsible for about two-thirds of a team's shooting percentage, according to a study by Tore Purdy.* So I also tried calculating each team's *expected* shooting percentage, based on the previous career shooting percentages of the individuals who took the shots, but that wasn't much better.[17] But stick with me because I'm going somewhere with this.

[* Tore Purdy (JLikens), "Team Even Strength Shooting Talent," *Objective NHL* (blog), May 29, 2011, http://objectivenhl.blogspot.ca/2011/05/even.html.]

In theory, it may be easier to face shots taken by Buffalo Sabres shooters than those attempted by Tampa Bay, but in practice there isn't enough variation in shooting percentages from team to team, and consequently from goalie to goalie. If you consider all 36 goalies who have faced at least 2,000 shots over the past three seasons combined, only the curious case of Anaheim's Frederik Andersen falls outside the narrow range of 8.5% to 8.7%.

---

17 Rob Vollman, "Questions and Answers," in Hockey Abstract 2014 (author, 2014), 300.

To me, it seemed almost impossible to separate the signal from the noise, but fortunately that didn't stop one prominent new blogger from trying. Dawson Sprigings integrated several aforementioned concepts to create expected save percentage (xSV%). The initial version we'll see first is "based on a 110 game moving average of the opposing shooter at the time of each shot faced by a goalie."*

[* Dawson Sprigings (DTMAboutHeart), "xSV%—Save Percentage Accounting for Shot Quality," *Don't Tell Me About Heart* (blog), March 13, 2015, http://donttellmeaboutheart.blogspot.ca/2015/03/xsvsave-percentage-accounting-for-shot.html.]

Based on this new perspective that takes shooter talent into account (as a proxy for shot speed and accuracy), Montreal's Carey Price remains the top challenger for Boston's Tuukka Rask's spot as the league's best puck stopper. Over this three-year period, Price faced 3,768 even-strength shots on which he was expected to allow 310.5 goals (xGA), based on the skill of the shooters, for an expected save percentage (xSV%) of 0.9176. Instead, he allowed just 246 goals (GA), for an actual save percentage (SV%) of 0.9347, which works out to a net difference of 0.0171. In absolute terms, Price has saved Montreal 64.5 goals relative to expectations (GS). This perspective puts Price, and Rask, safely ahead of the pack.

EXPECTED SAVE PERCENTAGE (XSV%) AT EVEN STRENGTH, 2012–13 TO 2014–15*

GOALIE	TEAM	SHOTS	XGA	XSV%	GA	SV%	DIFF	GS
Tuukka Rask	BOS	3,759	307.1	0.9183	242	0.9356	0.0173	65.1
Carey Price	MTL	3,768	310.5	0.9176	246	0.9347	0.0171	64.5
Henrik Lundqvist	NYR	3,462	281.5	0.9187	237	0.9315	0.0128	44.5
Sergei Bobrovsky	CBJ	3,416	274.8	0.9196	233	0.9318	0.0122	41.8
Corey Crawford	CHI	3,177	261.8	0.9176	224	0.9295	0.0119	37.8
Braden Holtby	WSH	3,586	292.3	0.9185	255	0.9289	0.0104	37.3
Craig Anderson	OTT	2,668	217.0	0.9187	181	0.9322	0.0135	36.0
Cory Schneider	VAN/NJD	3,022	247.1	0.9182	212	0.9298	0.0116	35.1
Steve Mason	PHI	3,035	250.4	0.9175	217	0.9285	0.0110	33.4
Jonas Hiller	ANA/ CGY	2,715	227.3	0.9163	195	0.9282	0.0119	32.3
Jonathan Bernier	LAK/TOR	2,968	241.6	0.9186	215	0.9276	0.0090	26.6
Ben Bishop	OTT/TBL	3,125	260.6	0.9166	234	0.9251	0.0085	26.6
Roberto Luongo	VAN/FLA	3,065	251.4	0.9180	225	0.9266	0.0086	26.4
Pekka Rinne	NSH	2,893	241.3	0.9166	215	0.9257	0.0091	26.3
Jimmy Howard	DET	3,072	249.0	0.9189	227	0.9261	0.0072	22.0
Cam Talbot	NYR	1,334	108.8	0.9184	88	0.9340	0.0156	20.8
Jonathan Quick	LAK	3,102	259.5	0.9163	239	0.9230	0.0066	20.5
Ryan Miller	BUF/STL/ VAN	3,386	287.9	0.9150	270	0.9203	0.0053	17.9
Semyon Varlamov	COL	3,784	309.8	0.9181	292	0.9228	0.0047	17.8

GOALIE	TEAM	SHOTS	XGA	XSV%	GA	SV%	DIFF	GS
Marc-André Fleury	PIT	3,466	282.8	0.9184	268	0.9227	0.0043	14.8
Antti Niemi	SJS	3,774	305.1	0.9192	291	0.9229	0.0037	14.1
Devan Dubnyk	EDM/ NSH/ ARI/MIN	2,864	233.9	0.9183	221	0.9228	0.0045	12.9
Brian Elliott	STL	1,873	155.9	0.9168	144	0.9231	0.0064	11.9
Michal Neuvirth	WSH/ BUF/NYI	1,473	124.7	0.9154	114	0.9226	0.0072	10.7
Mike Smith	ARI	3,671	304.2	0.9171	295	0.9196	0.0025	9.2
Jaroslav Halak	STL/ WSH/ NYI	2,604	213.3	0.9181	206	0.9209	0.0028	7.3
Thomas Greiss	SJS/ARI/ PIT	1,021	82.8	0.9189	76	0.9256	0.0067	6.8
Kari Lehtonen	DAL	3,717	303.8	0.9183	298	0.9198	0.0016	5.8
Jhonas Enroth	BUF/DAL	2,155	179.8	0.9166	174	0.9193	0.0027	5.8
Chad Johnson	ARI/BOS/ NYI	1,046	84.8	0.9189	79	0.9245	0.0056	5.8
Anton Khudobin	BOS/CAR	1,931	159.8	0.9173	154	0.9202	0.0030	5.8
Frederik Andersen	ANA	1,765	141.3	0.9199	136	0.9229	0.0030	5.3
Darcy Kuemper	MIN	1,271	106.3	0.9164	102	0.9197	0.0034	4.3
Eddie Lack	VAN	1,772	143.8	0.9188	140	0.9210	0.0022	3.8
Karri Ramo	CGY	1,583	132.0	0.9166	129	0.9185	0.0019	3.0
Ray Emery	CHI/PHI	1,485	124.8	0.9160	122	0.9178	0.0019	2.8
James Reimer	TOR	2,514	202.4	0.9195	203	0.9193	−0.0002	−0.6
Ilya Bryzgalov	PHI/ EDM/ MIN/ ANA	1,645	137.0	0.9167	140	0.9149	−0.0018	−3.0
Robin Lehner	OTT	1,804	147.7	0.9181	151	0.9163	−0.0018	−3.3
Al Montoya	WPG/ FLA	1,108	92.1	0.9169	96	0.9134	−0.0035	−3.9
Carter Hutton	NSH	1,196	98.9	0.9173	104	0.9130	−0.0043	−5.1
Reto Berra	CGY/ COL	1,000	84.6	0.9154	90	0.9100	−0.0054	−5.4
Curtis McElhinney	CBJ	1,279	104.5	0.9183	111	0.9132	−0.0051	−6.5
Evgeni Nabokov	NYI/TBL	1,961	161.8	0.9175	169	0.9138	−0.0037	−7.2
Martin Brodeur	NJD/STL	1,387	114.8	0.9172	123	0.9113	−0.0059	−8.2

GOALIE	TEAM	SHOTS	XGA	XSV%	GA	SV%	DIFF	GS
Justin Peters	CAR/ WSH	1,136	97.8	0.9139	106	0.9067	−0.0072	−8.2
Cam Ward	CAR	2,142	176.5	0.9176	185	0.9136	−0.0040	−8.5
Tim Thomas	FLA/DAL	1,087	88.3	0.9188	97	0.9108	−0.0080	−8.7
Anders Lindback	TBL/ DAL/BUF	1,590	133.5	0.9160	143	0.9101	−0.0060	−9.5
Ondrej Pavelec	WPG	3,294	269.0	0.9183	279	0.9153	−0.0030	−10.0
Viktor Fasth	ANA/ EDM	1,354	111.1	0.9180	123	0.9092	−0.0088	−11.9
Dan Ellis	CAR/ DAL/FLA	1,018	85.7	0.9158	98	0.9037	−0.0120	−12.3
Niklas Backstrom	MIN	1,633	134.7	0.9175	149	0.9088	−0.0088	−14.3
Ben Scrivens	TOR/ LAK/ EDM	2,635	211.1	0.9199	228	0.9135	−0.0064	−16.9

Note: Only goaltenders who faced a minimum of 1,000 shots are included.

[* Data obtained from *Don't Tell Me About Heart* (blog), http://donttellmeaboutheart.blogspot.ca/p/xsv.html.]

Just like with shot location, adjusting for the speed and accuracy of the shooter doesn't change the picture much, but it *does* change.

- Even though this table is ranked in terms of overall goals prevented, Price and Rask remain atop the list even when considering the data on a per-shot basis.
- The biggest surprise is Ottawa's Craig Anderson, who sits atop the tight pack of the next seven elite goalies.
- Cory Schneider and Braden Holtby remain in the group that's one step back, along with mainstays like Henrik Lundqvist and Sergei Bobrovsky.

- The pack is so tight that some goalies are in the picture when one factor is considered and just outside when it's another. In this case, Jonas Hiller, Corey Crawford, and Steve Mason are in.
- As usual, Cam Talbot is the best of the backups and platoon goalies, with Michal Neuvirth being the one to challenge Thomas Greiss for second.

High-Danger Situations

Even after accounting for shot location, which includes distance and angle, and the talent of the shooter, which implies the shot's speed and accuracy, there is still a component of shot quality that remains. Why? Because there are more players on the ice than just the shooter and the goalie, and there's a lot they can do to affect shot quality. Well, except the goaltender at the far end of the ice, I suppose.

Although the recorded effect appears minimal, players can improve their teammates' shooting percentages, most notably by setting up the plays and screening the goalies. That's why "forwards have a measurable bearing on their on-ice [shooting percentage] whereas defencemen typically don't," according to Domenic Galamini.* The effect isn't huge, however, as Matt Cane found that "while forwards do show more control of whether the puck goes in when they're on the ice, a single season's data is still about 75% luck for their own shots and 80% luck for their teammates."** Still, this is an area worth examining in our quest for the league's best puck stopper.

[* Domenic Galemini, "Possession Isn't Everything, with the Exception of Defensemen," *Own the Puck* (blog), December 8, 2014, http://ownthepuck.com/2014/12/08/possession-isnt-everything-with-theexception-of-defensemen/.]

[** Matt Cane, "How Much Skill Exists in On-Ice Shooting Percentages?," *Puck Plus Plus* (blog), December 11, 2014, http://puckplusplus.com/2014/12/11/how-much-skill-exists-in-on-ice-shooting-percentages/.]

When dealing with the most dangerous types of shots, rebounds are a good place to start. While they aren't explicitly recorded in NHL game files, rebounds can be inferred from the time difference in between shots. For example, Tom found that shots taken within two seconds of the previous attempt go in 28% of the time—the vast majority of these are likely rebounds.[18] Generating more rebounds is a good reason why teams can be successful by taking more shots, even low-probability attempts from the outside.

This temporal dimension of shot quality can be studied in even broader terms, beyond counting rebounds. For example, in an approach similar to what was done

---

18 Tom Awad, "Shot Quality," in Hockey Abstract 2014, by Rob Vollman (author, 2014), 35.

with shot distance, Fangda Li looked at the odds of each shot going in based on the elapsed time between shot attempts. Based on all regular-season data from 2008–09 through 2013–14, there's somewhat of a U-shape to the results.*

[* Fangda Li, "NHL Data Mining Part 1 – Shot Quality and Inter-arrival Times Between Corsi Events," *Pension Plan Puppets* (blog), October 16, 2014, http://www.pensionplanpuppets.com/2014/10/16/6986961/nhl-data-mining-part-1-shot-quality-and-inter-arrival-times-between.]

- It was confirmed that shot attempts that occur within four seconds of the previous one (i.e., rebounds) are up to four times more likely to go in.
- Shots taken between six and 11 seconds are actually less likely than average to score, despite being the most common.
- The odds of scoring begin to rise again as more time elapses.

Why does the shooting percentage start to rise when there's a long gap between shots? Because of the impact of breakaways and rushes, one would presume. In fact, breakaways are the only type of shot that's more dangerous than a rebound. While the NHL game files don't include exact data, one would have to assume that the odds of scoring on a breakaway would be comparable to penalty shots, which result in goals 31.5% of the time.

While breakaways can't be isolated, this temporal perspective can generally help identify which shots were taken off the rush. David Johnson defined them as any shot that occurred within 10 seconds of an event in the attacker's defensive zone, like a faceoff, a hit, a giveaway, a take-away, or a shot—basically anything that is recorded in an NHL game file.* Ultimately, that definition included 23.5% of all shots (but 28.6% of all goals).

[* David Johnson, "Introducing Rush Shots," *Hockey Analysis* (blog), July 9, 2014, http://hockeyanalysis.com/2014/07/09/introducing-rush-shots/.]

It makes intuitive sense that shots off the rush, especially in an odd-man situation, are far more dangerous than other shots. Applying this definition to data from 2007–08 through 2013–14 combined, Johnson found that "the league has a rush shot shooting percentage of 9.56% over the past 7 seasons while the shooting percentage is just 7.34% on shots we cannot conclusively define as a rush shot."*

[* David Johnson, "Introducing Rush Shots," *Hockey Analysis* (blog), July 9, 2014, http://hockeyanalysis.com/2014/07/09/introducing-rush-shots/.]

Based on this evidence, *War on Ice'* s aforementioned definitions of low-, medium-, and high-danger shots were modified to bump rebounds and rush shots up a level. That is, a rush shot from outside the expanded home-plate area is still considered a medium-risk shot, while a rebound from within the home-plate area is treated as a high-risk opportunity. In that sense, we have already accounted for the two most common types of dangerous shots.

Another prominent analyst on a quest to identify high-danger shots is former NHL netminder Steve Valiquette, who introduced his MSG Network viewers to the important concepts of the royal road and green, yellow, and red shots in 2014–15.*

[* Steve Valiquette, "Green Shots & Red Shots," *Vally's View* (blog), December 13, 2014, http://www.msg.com/shows/hockey-night-live/vally-s-view/green---red-shots.html.]

According to his manual tracking, 76% of goals are a result of green shots, which are attempts generated off pucks that were either carried or passed cross-ice (i.e., across the royal road, an imaginary line that extents from one net to the other), were one-timers, or involved screens, deflections, and/or broken plays.

The safest types of shots were termed red shots, which include wraparounds, net crashing, and unscreened shots off the rush. Based on his own experience and manually tracked data, Valiquette estimates that the odds of scoring are less than 3% if the goalie has at least half a second to see the shot.

Valiquette isn't the only one to manually track the data and to quantify the greater danger in certain kinds of shots. While a longstanding irritation of mine is how the NHL game files don't include any passing-related data whatsoever, Ryan Stimson has been conducting a project in which he and a group of independent bloggers and analysts have been recording such data themselves. And sure enough, they have found that shots taken immediately after a pass are more likely to go in.*

[* Ryan Stimson, "How Passing Relates to Shooting Percentage and Close Situations," *Hockey Prospectus* (blog), December 7, 2014, http://www.hockeyprospectus.com/stimson-how-passing-relates-toshooting-percentage-and-close-situations/.]

In a study of the five teams for which his team has recorded sufficient data, Stimson found that the shooting percentage was notably higher off a pass in every case and by about 5% on average. The

Panthers, for example, were scoring on over 13% of their shots off passes but less than 7% otherwise.

When all of this information is put together, it results in a brand-new version of Sprigings's expected save percentage.* This version still includes the speed and accuracy of the shooter, but it also incorporates shot distance, the type of shot, whether or not it was a rebound or on the rush, and even the current score (more on that coming up). Furthermore, the rolling average of the shooting percentage of the shooters has been increased to 275 shots for defencemen and 375 for forwards, which yields a far better estimate of their abilities to take hard and accurate shots.

[* Dawson Sprigings (DTMAboutHeart), "Updated xSV%—Save Percentage Accounting for Shot Quality," *Don't Tell Me About Heart* (blog), June 3, 2015, http://donttellmeaboutheart.blogspot.ca/2015/06/updated-xsv-save-percentage-accounting.html.]

Before revealing the results, I would be remiss if I didn't mention shot-quality neutral save percentage (SQN SV%), which was introduced by one of our hobby's true pioneers, Alan Ryder, long ago. It was essentially the exact same statistic, but it was built almost entirely independently and completely from scratch.* It's quite amazing that it took us over a decade to fight our way back to this and his other groundbreaking advances.

[* Alan Ryder, "Shot Quality," Hockey Analytics, 2004, accessed May 9, 2015, http://hockeyanalytics.com/Research_files/Shot_Quality.pdf.]

As for today's xSV%, the following leaderboard encapsulates everything that's been covered so far and represents our best guess at each goaltender's true abilities to stop the puck. Yes, after all this work and a couple dozen pages, we have finally arrived at our ultimate leaderboard. The leader? Montreal's Carey Price unseats Boston's Tuukka Rask as the league's best puck stopper, and by a comfortable margin over both him and the rest of the pack.

Over this three-year period, Price faced 3,784 even-strength shots of which he was expected to allow 290.1 goals (xGA), based on the quality of the shots faced, for an expected save percentage (xSV%) of 0.9233. Instead, he allowed just 246 goals (GA), for an actual save percentage (SV%) of 0.935, which works out to a net difference of 0.0117 in save percentage. In absolute terms, Price has saved Montreal 44.1 goals relative to expectations (xGS).

EXPECTED SAVE PERCENTAGE (XSV%) AT EVEN STRENGTH, 2012–13 TO 2014–15*

GOALIE	TEAM	SHOTS	XGA	XSV%	GA	SV%	DIFF	XGS
Carey Price	MTL	3,784	290.1	0.9233	246	0.9350	0.0117	44.1
Tuukka Rask	BOS	3,759	277.7	0.9261	242	0.9356	0.0095	35.7
Jonas Hiller	ANA/CGY	2,715	222.6	0.9180	195	0.9282	0.0102	27.6
Sergei Bobrovsky	CBJ	3,392	259.5	0.9235	233	0.9313	0.0078	26.5
Corey Crawford	CHI	3,177	250.0	0.9213	224	0.9295	0.0082	26.0
Cory Schneider	VAN/NJD	3,022	237.8	0.9213	212	0.9298	0.0085	25.8
Henrik Lundqvist	NYR	3,462	261.0	0.9246	237	0.9315	0.0069	24.0

GOALIE	TEAM	SHOTS	XGA	XSV%	GA	SV%	DIFF	XGS
Jonathan Bernier	LAK/TOR	2,968	237.6	0.9199	215	0.9276	0.0076	22.6
Craig Anderson	OTT	2,669	200.0	0.9251	181	0.9322	0.0071	19.0
Semyon Varlamov	COL	3,784	309.8	0.9181	292	0.9228	0.0047	17.8
Steve Mason	PHI	3,035	233.4	0.9231	217	0.9285	0.0054	16.4
Jimmy Howard	DET	3,072	239.4	0.9221	227	0.9261	0.0040	12.4
Cam Talbot	NYR	1,334	100.4	0.9247	88	0.9340	0.0093	12.4
Braden Holtby	WSH	3,586	265.4	0.9260	255	0.9289	0.0029	10.4
Ben Bishop	OTT/TBL	3,124	240.8	0.9229	233	0.9254	0.0025	7.8
Roberto Luongo	VAN/FLA	3,070	232.2	0.9244	226	0.9264	0.0020	6.2
Michal Neuvirth	WSH/ BUF/NYI	1,474	120.2	0.9185	114	0.9227	0.0042	6.2
Anton Khudobin	BOS/CAR	1,930	158.9	0.9177	153	0.9207	0.0030	5.9
Ryan Miller	BUF/STL/ VAN	3,386	275.4	0.9187	270	0.9203	0.0016	5.4
Marc-André Fleury	PIT	3,466	272.6	0.9213	268	0.9227	0.0013	4.6
Frederik Andersen	ANA	1,765	139.6	0.9209	136	0.9229	0.0020	3.6
Thomas Greiss	SJS/ARI/ PIT	1,021	79.1	0.9225	76	0.9256	0.0030	3.1
Eddie Lack	VAN	1,773	143.1	0.9193	140	0.9210	0.0017	3.1
Chad Johnson	ARI/BOS/ NYI	1,046	81.6	0.9220	79	0.9245	0.0025	2.6
Jhonas Enroth	BUF/DAL	2,155	176.4	0.9182	174	0.9193	0.0011	2.4
James Reimer	TOR	2,513	205.4	0.9183	203	0.9192	0.0010	2.4
Jaroslav Halak	STL/ WSH/NYI	2,604	207.7	0.9202	206	0.9209	0.0007	1.7
Devan Dubnyk	EDM/ NSH/ARI/ MIN	2,869	223.7	0.9220	222	0.9226	0.0006	1.7
Jonathan Quick	LAK	3,101	239.7	0.9227	239	0.9229	0.0002	0.7
Ray Emery	CHI/PHI	1,488	121.3	0.9185	122	0.9180	−0.0005	−0.7
Karri Ramo	CGY	1,583	127.4	0.9195	129	0.9185	−0.0010	−1.6
Pekka Rinne	NSH	2,899	214.3	0.9261	216	0.9255	−0.0006	−1.7
Kari Lehtonen	DAL	3,716	295.9	0.9204	298	0.9198	−0.0006	−2.1
Mike Smith	ARI	3,672	292.8	0.9202	295	0.9197	−0.0006	−2.2
Brian Elliott	STL	1,873	140.5	0.9250	144	0.9231	−0.0019	−3.5
Al Montoya	WPG/FLA	1,108	89.9	0.9189	96	0.9134	−0.0055	−6.1

GOALIE	TEAM	SHOTS	XGA	XSV%	GA	SV%	DIFF	XGS
Darcy Kuemper	MIN	1,271	95.8	0.9246	102	0.9197	−0.0049	−6.2
Reto Berra	CGY/COL	1,000	83.3	0.9167	90	0.9100	−0.0067	−6.7
Evgeni Nabokov	NYI/TBL	1,961	161.0	0.9179	169	0.9138	−0.0041	−8.0
Justin Peters	CAR/WSH	1,136	96.7	0.9149	105	0.9076	−0.0073	−8.3
Antti Niemi	SJS	3,774	281.3	0.9255	291	0.9229	−0.0026	−9.7
Robin Lehner	OTT	1,804	140.9	0.9219	151	0.9163	−0.0056	−10.1
Dan Ellis	CAR/ DAL/FLA	1,018	87.5	0.9141	98	0.9037	−0.0103	−10.5
Viktor Fasth	ANA/ EDM	1,354	112.3	0.9170	123	0.9092	−0.0079	−10.7
Curtis McElhinney	CBJ	1,283	99.4	0.9225	111	0.9135	−0.0090	−11.6
Ilya Bryzgalov	PHI/ EDM/ MIN/ANA	1,645	128.3	0.9220	140	0.9149	−0.0071	−11.7
Ondrej Pavelec	WPG	3,294	266.8	0.9190	279	0.9153	−0.0037	−12.2
Cam Ward	CAR	2,142	172.3	0.9195	185	0.9136	−0.0059	−12.7
Carter Hutton	NSH	1,196	91.0	0.9239	104	0.9130	−0.0109	−13.0
Tim Thomas	FLA/DAL	1,075	82.8	0.9229	96	0.9107	−0.0122	−13.2
Martin Brodeur	NJD/STL	1,387	107.1	0.9228	123	0.9113	−0.0115	−15.9
Ben Scrivens	TOR/ LAK/EDM	2,635	207.4	0.9213	228	0.9135	−0.0078	−20.6
Niklas Backstrom	MIN	1,633	127.9	0.9217	149	0.9088	−0.0129	−21.1
Anders Lindback	TBL/DAL/ BUF	1,590	121.3	0.9237	143	0.9101	−0.0136	−21.7

Note: Only goaltenders who faced a minimum of 1,000 shots are included.

[* Data obtained from "xSV%," *Don't Tell Me About Heart* (blog), http://donttellmeaboutheart.blogspot.ca/p/xsv.html.]

At first glance, this all may seem somewhat anti-climactic. Many readers may be wondering why they purchased this book to simply be told what the Vezina voters already knew. Others are sticking a finger on this page, flipping back to the overall save percentage at the beginning of the chapter, and wondering why they slogged through all this analysis to find out that their favourite goalie barely budged in the rankings. What gives?

Even after all this work adjusting for every factor of even remote significance, the leaderboard is not dramatically different from the starting point—but it *is* different. And, more importantly, there's a great confidence in having such a deep understanding of save percentage, its applications, and its limitations.

Maybe we already knew that Price and Rask were the best, but now we know *why.* And we certainly didn't know that Calgary's Jonas Hiller would rank third overall and second only to Price on a per-shot basis. And there are other interesting results:

- Once again, Cory Schneider is also included in that top tier of goalies, but Braden Holtby has fallen slightly out of the picture.
- Corey Crawford and Craig Anderson have remained with Henrik Lundqvist and Sergei Bobrovsky in that strong second level of goalies.
- Jonathan Bernier is also up and among that top pack of goalies.
- Steve Mason and Roberto Luongo have fallen out of the picture, Jaroslav Halak is barely above average, and Pekka Rinne has actually cost the Predators a goal or two.
- Among backups and platoon goalies, Cam Talbot remains on top but without a clear number two. In particular, Thomas Greiss is mixed in with Michal Neuvirth, Chad Johnson, and even Anton Khudobin.

Despite this leaderboard including virtually every scrap of information that can be gathered from NHL game files, it's important to remember just how significant of an impact random variation can have. I would never say that Craig Anderson is a better puck stopper than Roberto Luongo, for example, but I would say that the odds are in his favour.

Finally, there are a number of small but important factors that are more difficult to quantify and/or capture in the official score-sheet but that can nevertheless move the needle on this leaderboard. And I'm not just saying this to avoid alienating fans whose favourite goalie ranked poorly. It's really true.

Team Effects

A goalie's save percentage can be affected by the team for which he plays in ways that go beyond what can be captured in shot quality. That's why there's so much anecdotal evidence that playing for certain teams can give goalies such a huge advantage over others.

Take the Bruins, for example. Goalies have a surprising tendency to play like giants while in Beantown and to tumble to mediocrity (or below) when they leave town.

- Tim Thomas posted an overall 0.921 save percentage in seven full seasons in Boston but just

0.908 in his final split-season with Florida and Dallas.

- Anton Khudobin recorded a 0.926 save percentage with Boston but 0.914 over his first two seasons in Carolina.
- In 2013–14, Chad Johnson posted a nifty 0.925 save percentage in Boston, which fell to 0.889 his following season, with the Islanders.
- Remember Andrew Raycroft? In Boston he posted a 0.908 save percentage before being acquired by Toronto (Tuukka Rask—ouch), where his save percentage was a disappointing 0.890 over two seasons.
- Alex Auld, who had a career save percentage of 0.904 with eight different teams, enjoyed his best (half-)season in 2007–08 in Boston, with a 0.919 save percentage.
- Even depth option Hannu Toivonen, who posted a 0.878 save percentage in 23 games with St. Louis in 2007–08, previously posted a more respectable 0.896 mark in Boston.

This incredible set of results seems to go beyond what can be explained by random variation or the relatively limited impact of shot quality. It's only natural for history like this to call Tuukka Rask's numbers into question, not to mention opening up the possibility that other goalies are being similarly boosted (or penalized).

While the rather extreme situation in Boston is a bit of a puzzler, there are several team-related effects that haven't been captured so far. The different ways that data is recorded from city to city, the adjustments teams make late in the game based on the score, and various consequences of each team's roster and strategies can all have an impact on the save percentages of their goaltenders.

Recording Bias

One of the greatest frustrations in dealing with official NHL game files is the inconsistency with which the data is recorded. As Iain explained earlier, whether it's hits and blocks or shots and takeaways, the scorekeeping teams in some cities will count absolutely every single event while others will record only those about which they can be absolutely certain. That's why some of the observed differences in save percentages between teams are actually the result of recording bias and not the team's roster or strategy. Since every single goal is recorded but the number of shots that are recorded varies from rink to rink, save percentages can be affected by the arena in which goalies typically play, whether it's at home or against frequent divisional opponents.

For example, goalies like Los Angeles's Jonathan Quick, St. Louis's Brian Elliott, and Martin Brodeur, formerly of the New Jersey Devils, have all faced an unusually low volume of shots over recent years. To what extent was that a consequence of the team's talent and system versus a shot-recording bias in their home arena?

The most common way to get around problems like these is to use data for away games only, which should average out most of the various recording biases from city to city. However, there's hardly enough data with which to judge goalies as it is, so it doesn't make sense to throw half of it out. Instead, the stronger approach is to calculate the extent of this bias and adjust for it. That can be done by comparing the number of shots a team takes and allows in its home city relative to what the same team takes and faces elsewhere around the league.

Most recently, that's exactly what Matt Cane was inspired to do, when he noticed that there was virtually no correlation between goalie save percentages at home and on the road.* While it's natural to face a little more rubber on the road, Cane found that goalies like Quick, Minnesota's Niklas Backstrom, and Carolina's Cam Ward were facing up to three extra shots per road game, on average. While it's possible to explain those differences with random variation or if their team dramatically changes strategies on road trips, it's more likely that their shot totals at home are woefully under-counted. This could skew their own save percentages and those of the teams who frequently play against them in those home arenas.

[* Matt Cane, "Is Jonathan Quick Overrated? Shot Counting Bias and Save Percentages," *Puck Plus Plus* (blog), September 29, 2014, http://puckplusplus.com/2014/09/29/is-jonathan-quick-overrated-shotcounting-bias-and-save-percentages/.]

To address this, Cane made his own adjusted version of save percentage, given that we are woefully lacking in those, too. He tried to create a more accurate count of a goalie's actual shot volumes at home based on the shot volumes on the road. Due to random variation, there's not much of a relationship between this adjusted version of home and road save percentage, but it's much closer.

Ultimately, the whole point of Cane's exercise was to see how significantly this shot-recording bias could impact a goalie's save percentages. The results were quite alarming—a goalie in an extreme situation like Quick, for example, had an adjusted even-strength save percentage that jumped from 0.9257 to 0.9295 over Cane's three-year sample (2011–12 through 2013–14). That's quite significant, and it's enough to lift a middling goalie who has been overlooked right up among the tight pack at the top.

To make matters even worse, it's not just the number of shots that's being recorded differently but also where and when they took place. Every year, we laugh at examples of shots marked as wraparounds when they occurred 60 feet from the net. Unless they were taken by the 6-foot-9 Zdeno Chara, there was obviously a recording problem somewhere along the way. All jokes aside, these inaccuracies can have a serious impact on several components of shot quality, including shot location, and the identification of rush shots and rebounds.

While priority throughout this book is given to the very latest studies, Cane's recent analysis of recording bias wasn't as comprehensive as some of the work by our field's earlier pioneers. Given its importance, and the considerable frustration it has caused the world of hockey analytics, there have been almost countless studies that have attempted to thoroughly identify, quantify, and/or adjust for this shot-recording bias in various ways.

- Long-time hockey analytics pioneer Alan Ryder was the first to take note of this recording bias, back in 2007, but he concluded that "shot quality is not broken, just don't use it without understanding it."*

[* Alan Ryder, "Product Recall Notice for Shot Quality," Hockey Analytics, June 1, 2007, accessed May 9, 2015, http://hockeyanalytics.com/2007/06/product-recall-notice-for-shot-quality/.]

- In 2009, Tore Purdy was one of several renowned analysts to study this problem, going back 13 NHL seasons to compare the total of all shots taken (for and against) at home versus those on the road for all teams. He demonstrated how some team variations persisted year after year while others quickly vanished, and he also dispelled the notion that home/road shot variations were caused by different playing strategies.*

[* Tore Purdy (JLikens), "Home Recording Bias: Shots on Goal," *Objective NHL* (blog), March 20, 2009, http://objectivenhl.blogspot.ca/2009/03/in-previous-posts-it-was-shown-how-some.html.]

- As further evidence to Purdy's second point, Tom Awad found that home/road shot totals can vary by up to 5% for certain teams but without a budge in the goal totals.*

[* Tom Awad, "Numbers on Ice: A Castle Built on Sand," *Hockey Prospectus* (blog), November 12, 2009, http://www.hockeyprospectus.com/puck/article.php?articleid=351.]

- That same year, shot-quality expert Ken Krzywicki built the first real model that removed what he called "observer bias" from shot data.*

[* Ken Krzywicki, "Removing Observer Bias from Shot Distance," Hockey Analytics, September 1, 2009, accessed May 9, 2015, http://www.hockeyanalytics.com/Research_files/SQ-DistAdj-RS0809-Krzywicki.pdf]

- In 2011, Ben Wendorf put all of this data into colourful charts of three-year moving averages over the previous 10 seasons, which really served to spread the word throughout the growing hockey analytics community.*

[* Ben Wendorf (Bettman's Nightmare), "Shot Counting, in Pictures," *Arctic Ice Hockey* (blog), June 29, 2011, http://www.arcticicehockey.com/2011/6/29/2228337/shot-counting-in-pictures.]

- More recently, Wesley Yue looked at both the raw counts and the bias in shot location when he introduced the aforementioned LAEGP statistic in 2013.*

[* Wesley Yue, "LAEGAP Methodology," *Hockey Metrics* (blog), August 28, 2013, http://hockeymetrics.net/laegap-methodology/.]

- Finally, the definitive study on how recording bias affects all subjective NHL statistics, including shots, came out in 2014, courtesy of Michael Schuckers and Brian Macdonald, the latter of whom is now with the Florida Panthers.*

[* Michael Schuckers and Brian Macdonald, "Accounting for Rink Effects in the National Hockey League," Statistical Sports Consulting, December 2, 2014, http://arxiv.org/pdf/1412.1035v1.pdf.]

Based on all of this research, are shots over-counted in Boston and in a way that can affect the final leaderboard? Or have they simply been fortunate to consistently find really good and/or hot goaltenders?

It's actually really hard to answer that, even for smart cookies like these, who can tackle numbers as easily as others can tackle a bag of chips. Originally, Krzywicki did establish that "those in Boston over reported distance, giving the illusion of easier shots on goal,"* but his study was based on the 2008–09 season, and the separate, long-term studies by both Purdy and Wendorf have demonstrated that the recording biases can come and go. Those findings may explain the success of Thomas, Auld, and Toivonen, but they might not still apply today.

[* Ken Krzywicki, "Removing Observer Bias from Shot Distance," Hockey Analytics, September 1, 2009, accessed May 9, 2015, http://www.hockeyanalytics.com/Research_files/SQ-DistAdj-RS0809-Krzywicki.pdf.]

More recent studies, like Yue's, found that shot distances were actually being recorded fairly normally in Beantown these days, and Schuckers and Macdonald found that only Florida and St. Louis had statistically significant variations in shot counts, not Boston.

Rask may be off the hook, but I can't write the same for everybody else. There is a long-standing consensus in all of these studies, and it's bad news for Henrik Lundqvist. Madison Square Garden was called "the bane of NHL shot analysis"* by Gabriel Desjardins and "a statistical nightmare" by Schuckers.** Furthermore, Tom wrote that "40 feet is actually 30 feet if you're in Madison Square Garden,"*** and Yue more specifically found that "MSG, notorious for its inaccuracy, records each shot on average 4.6 feet closer to the net than it is."****

[* Gabriel Dejardins (Hawerchuk), "Clean Up Your Act, Madison Square Garden," *Artic Ice Hockey* (blog),

October 18, 2010, http://www.arcticicehockey.com/2010/10/18/1756880/clean-up-your-act-madison-square-garden.]

[** Michael Schuckers and Brian Macdonald, "Accounting for Rink Effects in the National Hockey League," Statistical Sports Consulting, December 2, 2014, http://arxiv.org/pdf/1412.1035v1.pdf.]

[*** Tom Awad, "Numbers on Ice: A Castle Built on Sand," *Hockey Prospectus* (blog), November 12, 2009, http://www.hockeyprospectus.com/puck/article.php?articleid=351.]

[**** Wesley Yue, "LAEGAP Methodology," *Hockey Metrics* (blog), August 28, 2013, http://hockeymetrics.net/laegap-methodology/.]

As difficult as it may be to have great confidence in the quality of shot-related data in some NHL arenas, it's absolutely impossible in Madison Square Garden. Hopefully some of these problems will disappear when technology is used to accurately and consistently record shots, and their location, without the need for subjective human judgment. No more 60-foot wraparounds.

Score Effects

Another detail that has been known to sabotage an otherwise sound analysis is score effects. As Tom covered in considerable detail in a dedicated chapter in *Hockey Abstract 2014,* teams that are trailing take more shots but at a slightly decreased shooting percentage while creating the reverse situation for the team in the lead.[19]

How does that affect our hunt for the game's best puck stopper? Essentially, it means that two otherwise identical goalies will have different save percentages, higher for the one who is usually in the lead and lower for the netminder on the weaker team. This phenomenon can also serve to exaggerate the difference between goalies, since "weaker goaltenders are more likely to be trailing in the third period, when they will face fewer but more dangerous shots," as Tom wrote.[20] While this appears to be a team effect, it's really a score effect.

The impact of score effects can be quantified by comparing team shooting percentages in different game situations. In this case, Tom found that a team's shooting percentage can range from 7.5% when the score is tied and/or a team is trailing all the way up to between 8.3% and 8.7% when a team is leading by one or more goals. This difference is even more exaggerated in the third period, where it leaps all the way to 9.1% with a two-goal lead.

---

19 Tom Awad, "What Are Score Effects?," in Hockey Abstract 2014, by Rob Vollman (author, 2014), 34.

20 Tom Awad, "What Makes Good Players Good, Part 3: Goaltenders," in Hockey Abstract 2014, by Rob Vollman (author, 2014), 77.

The conventional wisdom is that the improved shooting percentages in these late-game, close-score situations are because of improved shot quality. That is, the trailing team is pinching more and giving up more odd-man rushes. If that's true, then we can move on because we have already adjusted for shot quality. Unfortunately, Tom found that "if a team has the lead, its shots are more likely to go in; if a team is trailing, its shots are less likely to go in, even taking shot distance, rebounds and power plays into account."[21]

Tom isn't the only party pooper, as these results were confirmed by a recent study by Andrew Thomas. Even when the shots are broken down into low-, medium-, and high-danger opportunities, there is still a small but clear increase in all shooting percentages as teams achieve and strengthen their leads.*

[* Andrew Thomas, "The Road to WAR, Part 10: Modern Goaltending and Shooting," *War on Ice* (blog), April 3, 2015, http://blog.war-on-ice.com/the-road-to-war-part-10-modern-goaltending-and-shooting/.]

SHOOTING PERCENTAGE BY SHOT DANGER, 2008–09 TO 2013–14*

---

21 Tom Awad, "What Makes Good Players Good, Part 3: Goaltenders," in Hockey Abstract 2014, by Rob Vollman (author, 2014), 77.

SCORE	LOW DANGER	MEDIUM DANGER	HIGH DANGER
Down 3+	3.22%	7.63%	16.91%
Down 2	3.17%	7.78%	18.52%
Down 1	3.20%	7.96%	18.22%
Tied	3.10%	7.72%	17.78%
Up 1	3.35%	8.22%	19.10%
Up 2	3.35%	8.88%	19.58%
Up 3+	3.42%	8.78%	19.43%

[* Data obtained from War on Ice (blog), http://blog.war-on-ice.com/the-road-to-war-part-10-moderngoaltending-and-shooting/.]

These results mean that goalies on weaker teams are going to have slightly lower save percentages, even after shot quality has been accounted for. Normally that would mean going back and applying some extra credit in our own analysis for puck stoppers on weaker teams, but Sprigings already included an adjustment for score effects in his latest version of xSV%. That means that goalies like New Jersey's Cory Schneider and Toronto's Jonathan Bernier are already being ranked properly on our leaderboard. So, we *can* move on.

Roster and System

Given that teams can influence shooting and save percentages in different score situations based on the players they send over the boards and/or the strategy they're tasked to employ, it stands to reason that some teams might tilt one way or the other all the time. That is, there could be some teams with an abundance of talent who play that aggressive, come-from-behind fashion all the time, just as there are likely teams with rosters composed primarily of those who play a conservative, lead-protecting way throughout the entire contest. If so, then that's an actual team effect on save percentage that may have slipped through the cracks.

There is not a great deal of consensus within the hockey analytics community about the extent to which teams can affect their goalie's save percentage, if at all—especially beyond what has been captured by recording bias and/or score effects. Even though it makes intuitive sense that there are certain players who can force shots to the outside and otherwise reduce the speed and accuracy of the shots that reach the net, there just hasn't been sufficient evidence—yet.

If teams had a significant impact on a goalie's save percentage, then it could be captured in the data when the netminders changed teams or by comparing starters to their backups. Years ago, smarter men than I searched for such evidence, but leading analysts like Phil Myrland, Vic Ferrari, and Tore Purdy came up short of anything that amounts to more than a goal or two per season.*

[* Phil Myrland, "Looking for Outliers," *Brodeur Is a Fraud* (blog), May 12, 2009, http://brodeurisafraud.blogspot.ca/2009/05/looking-for-outliers.html; Timothy Barnes (Vic Ferrari), "The Shot Quality Fantasy," *Irreverent Oiler Fans* (blog), July 24, 2009, http://vh

ockey.blogspot.ca/2009/07/shot-quality-fantasy.html; Tore Purdy (JLikens), "Team Effects and Even Strength Save Percentage," *Objective NHL* (blog), May 10, 2011, http://objectivenhl.blogspot.ca/2011/05/team-effects-and-even-strength-save.html.]

Quite frankly, these results just don't sit well with most fans. Given how it's already been demonstrated that players can affect their team's shooting percentages, presumably with great passing and positioning, not to mention the superior speed and accuracy of their own shots, it only makes sense that there should be individuals with the same talent but on the defensive side of the spectrum.

What can I say? Domenic Galamini went looking for evidence that players can have a persistent impact on their team's save percentages by comparing three-year samples of 2007–08 through 2010–11 to 2011–12 through 2013–14, and he found that "forwards and defencemen have virtually no control over on-ice save percentages."* Garret Hohl pursued the same goal and found that "there is no observable difference in successfully improving save percentage."**

[* Domenic Galamini, "Possession Isn't Everything, with the Exception of Defensemen," *Own the Puck* (blog), December 8, 2014, http://ownthepuck.com/2014/12/08/possession-isnt-everything-with-theexception-of-defensemen/.]

[** Garret Hohl, "Hockey Talk: On Player Control Over Save Percentage," *Hockey Graphs* (blog), April 6, 2015, http://hockey-graphs.com/2015/04/06/hockey-talk-on-player-control-over-save-percentage/.]

In my view, there are things that teams and their players do that affect save percentages, but potentially not in a way that can be definitively captured by statistics. Consider teams that block a lot of shots, for example. These teams allow fewer shots but may be disproportionately filtering out the less dangerous shots from far out, which will hurt the team's save percentage. In effect, teams that block shots will increase the shot quality that their goaltenders face, as Purdy first demonstrated years ago (although there is no consensus).*

[* Tore Purdy (JLikens), "Shotblocking and Save Percentage," *Objective NHL* (blog), November 19, 2008, http://objectivenhl.blogspot.ca/2008/11/shotblocking-and-save-percentage.html; CanesAndBluesFan, "Blocking Shots and Save Percentage," *St. Louis Game Time* (blog), July 14, 2014,http://www.stlouisgametime.com/2014/7/14/5900113/blocking-shots-and-save-percentage.]

Purdy also discovered that shot blocking has an additional impact on save percentages that goes beyond what can be accounted for by improved shot quality. Blocking shots increases the danger of the remaining shots by screening the goalies and/or causing accidental deflections. This drops save percentages even further, and it explains why teams that block more shots tend to have lower save percentages.

This isn't to argue that blocking shots is a bad strategy—far from it. Even if a failed block increases the chances of that shot going in, a successful block erases the chances of scoring completely. Even in a ludicrously extreme hypothetical scenario where a failed block will double the chances of a shot going in, Matt Cane calculated that a team should still attempt to block it if their chances of success are at least 60%.* In more realistic scenarios where a failed block only slightly increases the probability of scoring, the break-even point to make the attempt worthwhile is very low and almost always makes sense.

[* Matt Cane, "When Should Teams Attempt to Block Shots?," *Puck Plus Plus* (blog), August 26, 2014, http://puckplusplus.com/2014/08/26/when-should-teams-attempt-to-block-shots-looking-at-the-breakeven-success-rate-for-shot-blocking/.]

Whether or not blocking shots makes sense at the team level, it's the impact to the goalie's save percentage that's of consequence in our quest for the league's best puck stopper, and shot blocking is an example of how a team strategy can affect the results beyond what has been captured already.

To close with one more example, consider those teams that are now known to be worse at preventing clean zone entries, thanks to the data manually gathered by Corey Sznajder and others in the hockey analytics community. Since we know, both intuitively and statistically, that shots following clean zone entries are more dangerous than those following a dump and chase, these weak defensive teams must be hurting the save percentages of their goaltenders to a degree similar to the shot-blockers.

Even though its existence is difficult to prove, is without consensus, and has a net effect that is likely quite minor, team rosters and strategies likely have an impact on save percentages that should be kept in mind.

Workload

Now we're really starting to dig deep. While workload-related factors can theoretically have an impact on save percentages, there's a distinct lack of evidence. In fact, there's a real debate about whether or not the volume of shots goalies face really matters. Furthermore, it doesn't appear to matter if goalies get into a rhythm (at least not in a way that can be captured with statistics), and it turns out that the once-held belief that goalies shouldn't play back-to-back games might have been just an accidental fluke with the data.

Of all workload-related matters, the most contentious question is whether or not the average volume of shots a goalie faces will affect his save percentage. While this is certainly true on an individual game-by-game basis, there's no consensus on whether this is still true at season's end.

It's admittedly quite hard to examine this issue without tripping over recording bias, selection bias, score effects, and many of the other previously mentioned factors, which is why so many analysts land on each side of the issue.

- David Johnson wrote that "there appears to be a correlation between [shot attempts] and save percentage" that is "stronger if we restrict to goalies that haven't changed teams."*

[* David Johnson, "Does Higher Corsi Against Rates Boost Save Percentage?," *Hockey Analysis* (blog), November 24, 2014, http://hockeyanalysis.com/2014/11/24/higher-corsi-rates-boost-save-percentage/.]

- Likewise, Shane Beirnes found that "there is a clear relationship between save percentage and shot volume. This relationship appears on both a season-by-season and game-by-game basis," but he was also careful to add that this "does not mean that shots *cause* higher save percentages."*

[* Shane Beirnes, "Bernier's Save Percentage II," *Shane's Site* (blog), February 22, 2014, http://shanebeirnes.wordpress.com/2014/02/22/berniers-save-percentage-ii/.]

- Matt Cane is somewhat in the middle, finding that "in smaller sample sizes an individual goalie's save percentage could be skewed by the number of shots he faces. Over the long run this should even itself out for most goalies."*

[* Matt Cane, "Shots Against and Even Strength Shooting Percentage, *Puck Plus Plus* (blog), June 29, 2013, http://puckplusplus.com/2013/06/29/shots-against-and-even-strength-shooting-percentage/.]

- Greg Balloch was also somewhere on the fence, but he did conclude that, if it exists, "shot attempt quantity doesn't seem to be as big of a negative factor on save percentage as originally thought."*

[* Greg Balloch, "Goaltender Workload and Its Effect on Save Percentage," *In Goal Magazine* (blog), February 21, 2015, http://ingoalmag.com/analysis/goaltender-workload-and-its-effect-on-save-percentage/.]

- On the other hand, Garret Hohl found that "there is no substantial relationship in [goalies] playing better—in terms of save percentage—when facing more or less shots against."*

[* Garret Hohl, "Save Percentage vs the Experts: Do Shots Against Inflate a Goaltender's Save Percentage?," *Hockey Graphs* (blog), August 28, 2014, http://hockey-graphs.com/2014/08/28/savepercentage-vs-the-experts-do-shots-against-inflate-a-goaltenders-save-percentage/.]

- Dawson Sprigings also found that "there is still little to no evidence relating [shot attempts] a goaltender faces and their save percentage."*

[* Dawson Sprigings (DTMAboutHeart), "Corsi Against Doesn't Correlate with Save Percentage," *Don't Tell Me About Heart* (blog), November 27, 2014, http://donttellmeaboutheart.blogspot.ca/2014/11/corsi-against-doesnt-correlate-with.html.]

- Similarly, Braden Thompson found that "there is no correlation between save percentage and shot totals. When it comes down to it, there is no reason to believe goalies will be more effective if they face higher shot totals."*

[* Braden Thompson, "The Effect of Shot Totals on Save Percentage," *On the Forecheck* (blog), October 7, 2014, http://www.ontheforecheck.com/2014/10/7/6932789/effect-shot-totals-on-save-percentage.]

- The meticulous Tore Purdy concluded that "there is absolutely no evidence that high shot totals have an inflationary effect on goaltender save percentage."*

[* Tore Purdy (JLikens), "Save Percentage and Shots Against," *Objective NHL* (blog), October 22, 2008, http://objectivenhl.blogspot.ca/2008/10/goaltending-save-percentage-and-shot.html.] [* Tore Purdy (JLikens), "Save Percentage and Shots Against," *Objective NHL* (blog), October 22, 2008, http://objectivenhl.blogspot.ca/2008/10/goaltending-save-percentage-and-shot.html.]

- Goaltending analytics pioneer Phil Myrland thinks that "special teams, playing to the score, and strength of opposition effects are mainly responsible for the stats that show a relationship between shots against and save percentage."*

[* Phil Myrland, "Shots and Save Percentage Revisited," *Brodeur Is a Fraud* (blog), January 15, 2009, http://brodeurisafraud.blogspot.ca/2009/01/shots-and-save-percentage-revisited.html.]

- In the *Hockey Compendium,* early pioneers Jeff Klein and Karl-Eric Reif invented the goaltender perseverance stat based on the opposite premise, that it was actually *more difficult* to face higher volumes of shots.[22]
- As for me, I looked at this question in great detail in our last book and found that "the goalies who face relatively fewer shots ... fare no better nor worse than those who routinely get shelled."[23]

Of course, hockey analytics isn't a democracy or a popularity contest. If something is true, then it really doesn't matter how many people believe it isn't. The point of listing all of these studies is if shot volumes actually had a non-trivial impact on save percentages,

22 Jeff Z. Klein and Karl-Eric Reif, The Hockey Compendium: NHL Facts, Stats, and Stories (Toronto: McClelland & Stewart, 2001).

23 Rob Vollman, "Goaltending Analytics Revisited," in Hockey Abstract 2014 (author, 2014), 102.

then it probably shouldn't be so hard for all these great minds to prove (especially that last genius).

Part of the reasoning behind the theory that higher shot volumes will boost shooting percentage is the conventional wisdom that goalies get into a rhythm. If a goalie doesn't face enough shots, he can have difficulty staying warm and flexible and keeping his mind in the game—or at least that's how the argument goes.

That's certainly a popular narrative whenever a goalie lets in a bad goal after a long shotless stretch, but is it true? No, it isn't. There was a great article by Tyler Dellow that established that save percentages don't drop with every minute that elapses without a shot, but his website was taken down after he was hired by the Edmonton Oilers (which is not exactly a team that has had a lot of long shotless stretches).*

[* Tyler Dellow, "Does Save Percentage Fall as Time Elapses?," *MC79 Hockey* (blog), February 21, 2006, http://www.mc79hockey.com/?p=2447.]

Eric Tulsky, whose work fortunately remained available after his hiring by the Carolina Hurricanes, took it a step further, with a study of his own on the six-year anniversary of Dellow's study. This time he looked at whether or not goalies performed better when they had the opportunity to handle a few games in a row, and he found that "there is no evidence whatsoever that a lot of consecutive games will help a goalie get in a rhythm."*

[* Eric Tulsky, "Do Goaltenders Really Get in a Rhythm?," *Broad Street Hockey* (blog), February 21, 2012, http://www.broadstreethockey.com/2012/2/21/2814369/goalie-consecutive-starts-sorry-no-url-win-today.]

What about playing back-to-back games? Tulsky's *Broad Street Hockey* colleague Kurt discovered something intriguing in the lockout-shortened 2012–13 NHL season—that save percentages drop when a goalie plays consecutive games without a day of rest. Now that's interesting.

DAYS OF REST AND SAVE PERCENTAGE, 2012–13*

DAYS OF REST	SAVE %
0	0.892
1	0.913
2	0.910
3	0.911
4	0.914
5+	0.910

[* Kurt R, "Goalies on Back-to-Backs, or Why Steve Mason Should Start Tonight's Game Against the Maple Leafs," *Broad Street Hockey* (blog), April 4, 2013, http://www.broadstreethockey.com/2013/4/4/4176206/ilya-bryzgalov-steve-mason-michael-leighton-goalies-back-to-back.]

To eliminate the possibility that the low save percentage on back-to-back games was due to the goaltender himself and not poor defensive play from the equally fatigued team in front of him, Tulsky compared these results to games where a fresh goalie played the second part of those back-to-back games instead. Since the rested goalie had a save percentage of 0.912, it appeared at the time that getting more starts on zero days' rest would hurt a goalie's save percentage.*

[* Eric Tulsky, "Why Goalies Should Almost Never Start Back to Back Games," *Broad Street Hockey* (blog), April 8, 2013, http://www.broadstreethockey.com/2013/4/8/4188626/philadelphia-flyersgoaltending-stats-back-to-back-games.]

However, it turns out that these results were just a fluke of that season. The time lost to the lockout severely compressed the schedule, resulting in more back-to-back games while forcing teams to ride their starting goalies a lot harder than they would have over a longer season. That's why a blogger known as CanesAndBluesFan noticed stark differences between the data for the 2011–12 season and that of 2012–13 while observing that the confidence intervals overlapped, which means that you couldn't definitively say, statistically, that rested goalies performed better than those who were fatigued.*

[* CanesAndBluesFan, "On Why NHL Teams Should Never Play a Tired Goalie," *St. Louis Game Time* (blog), November 12, 2014, http://www.stlouisgametime.com/2014/11/12/7211791/on-why-nhl-teamsshould-never-play-a-tired-goalie.]

More recently, Andrew Thomas re-ran the exact same study, but for every season since the 2005 lockout, and he confirmed that the two seasons Tulsky studied just happened to be the worst on record. In fact, in three of the post-lockout seasons, goaltending save percentages were actually *higher* on zero days' rest.* Just as analysts like Phil Myrland and Jonathan Willis had previously established, most goalies don't significantly wear down, especially once the data is adjusted for factors such as home-ice advantage, shot location, score effects, manpower situation, and selection bias.** Those who do wear down are rarely placed in such situations anyway.

[* Andrew Thomas, "Back-to-Backs and Goalie Performance," *War on Ice* (blog), April 11, 2015, http://blog.war-on-ice.com/back-to-backs-and-goalie-performance/.]

[** Phil Myrland, "Goalie Fatigue," *Brodeur Is a Fraud* (blog), August 19, 2008, http://brodeurisafraud.blogspot.ca/2008/08/goalie-fatigue.html; Jonathan Willis, "Do Goalies Wear Down from Playing Too Many Games?," The Score, September 14, 2009, accessed May 9, 2015, http://blogs.thescore.com/nhl/2009/09/14/do-goalies-wear-down-from-playing-too-many-games/.]

While playing the tail end of a back-to-back may not significantly affect a goalie's performance in general, the same isn't true of the team playing in front of him. In the 2014 edition of *Hockey Abstract,* Tom found that "over the last five seasons, the overall save percentage of teams on the road playing the second night of back-to-back games was 90.84%, lower than the 91.08% they achieved when they were not fatigued."[24] Furthermore, Andrew Thomas found that "teams on the second day of a back-to-back are on average 5% worse at attempting close shots (including close tip-ins or deflections), and teams on the second day of an away-after-away are also 6% worse at allowing close shots."*

[* Andrew Thomas, "The Road to WAR (for Hockey), Part 3: Shot Quality Assurance, Plus a Bonus on Travel Fatigue," *War on Ice* (blog), October 5, 2014, http://blog.war-on-ice.com/the-road-to-war-forhockey-part-3-shot-quality-assurance-plus-a-bonus-on-travel-fatigue/.]

In essence, that means that the fatigued teams themselves have a little more difficulty scoring, and preventing scoring, on the second game of a back-to-back. Who does this affect? Tom found that backup goalies are almost twice as likely to get the difficult starts on the tail end of a road back-to-back and almost half of the fatigued home starts as well, so it's important to note but nothing that should cause a lot of shuffling among the league's starters.

---

24 Tom Awad, "What Makes Good Players Good, Part 3: Goaltenders," in Hockey Abstract 2014, by Rob Vollman (author, 2014), 78.

Other Factors

Even after all of this exhausting work, there are still a number of small factors that can affect a goalie's save percentage that don't fit into any of these buckets. None of them is worth writing home about, but then again, it's not like you would literally pen a letter to your mother about hockey analytics either.

- Due primarily to having the last line change, even-strength shooting percentages at home increase from 7.4% to 7.9%. As Tom discovered in *Hockey Abstract 2014,* this mostly affects backup goalies, who have smaller and more skewable sample sizes and who play 56% of their games on the road, compared to 46% for starters.[25] In extreme cases, like in Montreal or Colorado, the backup virtually never plays at home.
- Tom also found that shooting percentages, which start at 7.4% in the first period, leap to 8.2% in the second period, with the long change, and then go back down to 7.9% in the third period.[26] Of course, I can't imagine there are any goalies with an imbalance in the periods that they have played.

---

25 Tom Awad, "What Makes Good Players Good, Part 3: Goaltenders," in Hockey Abstract 2014, by Rob Vollman (author, 2014), 77.

26 Tom Awad, "What Are Score Effects?," in Hockey Abstract 2014, by Rob Vollman (author, 2014), 47.

- Even though there isn't any evidence that goalies perform better than usual after a bad start, there are still those pundits who believe that they're more likely to bounce back and really shine. As I discovered in the last *Hockey Abstract,* "there doesn't appear to be any evidence that goalies ... play any better than usual after a really bad start. In fact, it seems that at least a portion of whatever caused the bad game in the first place tends to continue."[27]

- Michael Lopez discovered that Sundays are lower scoring, oddly enough.* I doubt there are any goalies who play enough Sundays to skew their overall numbers, but this just goes to show that there are still so many other factors outside what's already been covered.

[* Michael Lopez, "Sunday Is a Day for Relaxing. Just Ask the NHL," *Stats by Lopez* (blog), February 9, 2009, http://statsbylopez.com/2015/02/09/sunday-is-a-day-for-relaxing-just-ask-the-nhl/.]

- There's a reason why teams look at the handedness of their defencemen. Matt Cane found that it can have an impact on shooting percentages based on where the shot was taken.* Anecdotally, I also remember a passionate conversation between Darren Pang and Craig Simpson, explaining to me

---

27 Rob Vollman, "Goaltending Analytics Revisited," in Hockey Abstract 2014 (author, 2014), 104.

how right-handed shooters have an advantage over left-handed catchers. Simpson, whose career 23.7% shooting percentage is third all-time, remarked how he would visualize the way he would approach opposing goalies based on their glovedness, while Pang expressed his great surprise that there were actually NHL goaltenders who went into a game unaware of the handedness of their opponents.

[* Matt Cane, "Shot Location Data and Strategy I: Off-Hand Defencemen," *Puck Plus Plus* (blog), November 11, 2014, http://puckplusplus.com/2014/11/11/shot-location-data-and-strategy-i-off-hand defencemen/.]

These examples were not meant to drop your jaw, simply to show that there are still a lot of factors, if you care to keep drilling down. We will never arrive at a perfect version of save percentage that includes only a goalie's true talent, but it's not from lack of effort.

Closing Thoughts

In our first book I wrote a chapter entitled "Who Is the Best Goalie?" in which I concluded that "the strongest case can be made for Henrik Lundqvist."[28] That may still be the case, but the evidence now

28 Rob Vollman, "Who Is the Best Goalie?," in Hockey Abstract (author, 2014), 52.

builds a more compelling case for Montreal's Carey Price or possibly Boston's Tuukka Rask.

Lundqvist nevertheless remains in a tight pack with three other up-and-coming goaltenders, Cory Schneider, Sergei Bobrovsky, and Braden Holtby, along with some combination of Jonas Hiller, Craig Anderson, Jonathan Bernier, Corey Crawford, and Steve Mason.

Why so much uncertainty? Is it to hedge my bets or maximize book sales by NHL market? No. It's because there is such great parity among the skilled and fortunate few who battle their way into scarce NHL opportunities that the dramatic effect caused by random variation, shot quality, and various team effects can completely obscure the picture.

It's possible that some of the upcoming technological advances will clear away some of the mystery, but the very nature of save percentage won't change, nor will the process by which the league's best puck stopper can be found. Maybe some day it will be obvious which of these 11 goalies is best, but right now the best guess is either Price or Rask.

EVERYTHING YOU EVER WANTED TO KNOW ABOUT SHOT-BASED METRICS (BUT WERE AFRAID TO ASK)

by TOM AWAD

There has been a revolution in hockey statistics over the last seven years; new statistics are appearing and being used and refined faster than many of us can keep up with. However, if you had to point to *one* development that was the turning point that jump-started the new hockey analytics movement, if would be the adoption and widespread usage of shot-based metrics. Shot-based, or "possession," metrics (the two terms are often used interchangeably) have become one of the primary ways of analyzing individual and team performances.

Many intelligent people have written about the importance and different aspects of shot-based metrics. My objective is not to claim any of their innovations as my own but simply to synthesize the state of the knowledge, so that someone familiar with hockey but with little knowledge of advanced stats can understand what they are, how they can be used, and what their limitations are.

From Plus/Minus to SAT

I will begin with a simple, uncontroversial statement: the objective of a hockey game is to outscore the opponent. To this end, each hockey player's job is to help his team outscore the opposition by as much as possible. This is fairly obvious, but it tells us to a large extent how hockey players' contributions should be measured and quantified—that is, by how much they contribute to outscoring the other team. Obviously each player has a different job, and his contributions should be measured in light of this: a goaltender whose team is trailing can't really help his team even the score, but he can help them from falling behind even further.

There are caveats to this rule, and everything depends on context. As we get closer to the end of a game, for a team that is trailing, goals for matter more than goals against. This is why, at the very end, teams pull their goaltender, because past a certain point goals against don't matter *at all.* The opposite is true for teams that are in the lead; late in the game we will often see them going into a defensive shell, simply clearing the puck out of the defensive zone, and in the extreme even icing it to reduce pressure from the other team. When protecting a one-goal lead late in the game, goals for don't matter *at all.*

However, for the vast majority of a hockey game, both goals for and goals against matter, and a player's job is to tilt that difference in his team's favour. This simple yet important insight led to the development of hockey's least understood and most maligned statistic, plus/minus.

The Beauty of Plus/Minus

My personal opinion is that plus/minus is very much like communism: extremely elegant in theory but not very good in practice.

Plus/minus was developed by the Montreal Canadiens in the 1950s, following the simple insight that what mattered about a player was whether he helped his team outscore the opponent. The concept of plus/minus is extremely basic: you get a +1 whenever your team scores a goal and you are on the ice and a −1 whenever the other team scores a goal and you are on the ice. The beauty of this system is that it doesn't require a coach or scorer to make a judgment call about which players participated in the goal or which ones are responsible for the other team getting a scoring chance and then capitalizing on it; all players are held equally responsible.

The idea behind plus/minus is that while a single player may be more responsible than the others on any individual goal, in the long run it will balance out. If your team is constantly being scored on while you're on the ice, you're probably doing something wrong, even if nobody can put their exact finger on it. In the same way, if your team is constantly scoring goals while you play, you are probably helping in some way, even if you don't get credit for a goal or an assist on every puck that goes in. Most importantly of all, by subtracting the two, plus/minus says if you're helping *more* than you're hurting, neatly encapsulating your contribution to your team in one easy-to-understand number.

The NHL started keeping plus/minus officially in 1967, and for a while it seemed to be doing what it was intended to do: showing which players were dominating the opposition while they were on the ice. For example,

- Bobby Orr led the league in plus/minus six times in seven years and posted a league record +124 in 1970–71, a record that hasn't been broken since and never will be;
- Larry Robinson has the second-highest mark of all time, at +120 in 1976–77, and his career +730 is another record that will likely never be beaten;
- Wayne Gretzky led the league four times and was a combined +551 in nine seasons with the Edmonton Oilers.

And then Gretzky was traded, the league stopped having dynasties, and plus/minus kind of stopped working. No, seriously. Look at this graph:

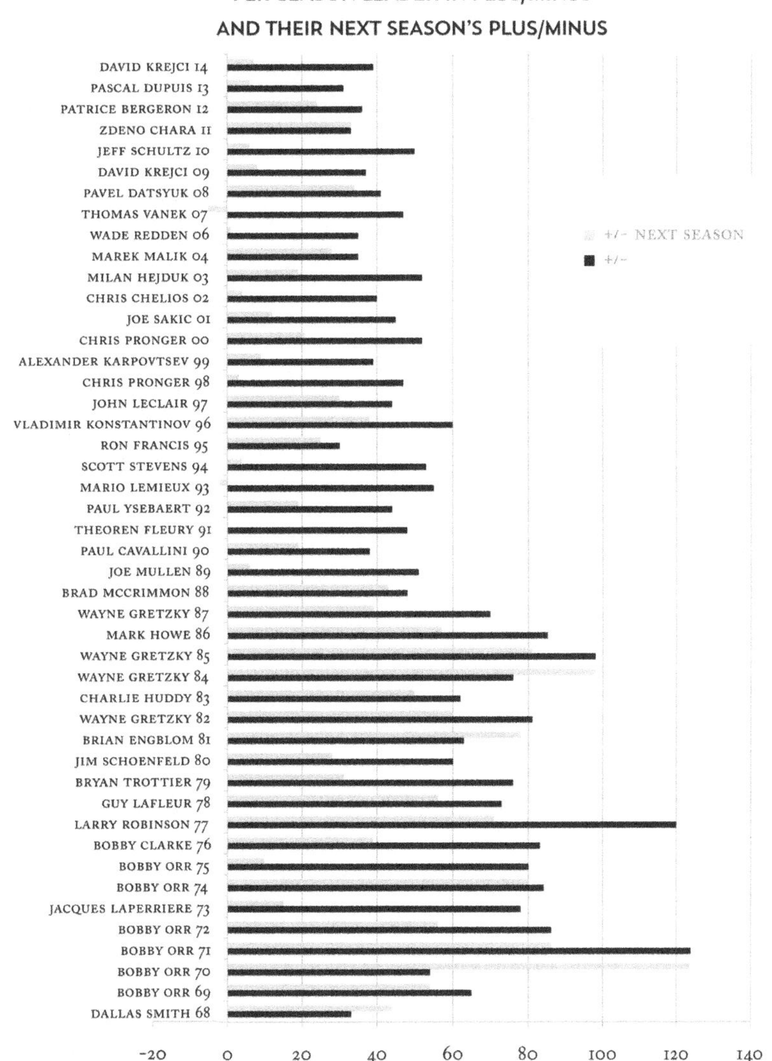

This is a fairly simple graph that shows, for each season, the league leader in plus/minus as well as how well that player did the next season. We can see that, prior to 1988, the league leader tended to have a pretty good plus/minus in the following season as well. In fact, the following season he averaged about 75% of his previous plus/minus and in several cases actually improved. One necessary (but not sufficient)

sign of a good statistic is that whatever it measures is somewhat repeatable; if it's not repeatable, then it's probably random, and if it's random, then we don't really care about it.

However, since 1989, the league leader in plus/minus has often had trouble repeating his success. Indeed, the leaderboard even has some curious members:

- Paul Ysebaert, who led the league with +44 in 1991–92, was +19 the next season, but he was −75 over the rest of his career and never did better than +3 in any other season.
- Thomas Vanek, never known as a defensive wizard, posted a +47 in 2006–07, was −5 the next year, and was −18 over the rest of his career with the Sabres.

On average, the league leader in plus/minus since 1988–89 has averaged only 32% of his mark the following year. To be fair, some of this has been the absence of generational talents such as Orr and Gretzky. Some of it is the absence of dynasties; between 1968 and 1985, all but three of the league leaders played for the Boston Bruins of the 70s, the Montreal Canadiens of 1973 to 1981, or the Edmonton Oilers of the 80s. But the real difference is that, in an age of parity, plus/minus's flaws finally caught up to it. Plus/minus remains theoretically perfect, but in practice it is almost useless.

The Six Problems with Plus/Minus

Plus/minus has six major flaws. They are important to know because this statistic shares three of them with its cousin, shot attempts (SAT), which we'll get to shortly.

1. CONSTRUCTION

The NHL made a significant error in the very definition of plus/minus when it included short-handed goals in the statistic. The NHL realized, correctly, that counting power-play goals was completely invalid; however, it didn't realize that including short-handed goals in the number also biased the statistic in a significant way. As an example, Alexander Ovechkin had the league's third-worst plus/minus in 2013–14, at −35. However, he was on the ice for nine shorthanded goals against, which counted against his plus/minus. Now −26 is not spectacular, but it's certainly better than −35. More importantly, the nine short-handed goals have *nothing to do* with how much Ovechkin helped or hindered his team at even strength; he is being penalized for playing on the power play, which he did extremely well. It is not plus/minus's fault that it has been miscalculated from the beginning, but it has always been a problem with the NHL's official plus/minus stats.

2. TEAM EFFECTS

By far, plus/minus's biggest flaw is that it is heavily influenced by which team a player skates for. The fourth-best plus/minus in history is Dallas Smith's +94 from 1970–71, and I doubt I need to convince you that, as good as Smith was, he wasn't the fourth-best player in the NHL in the last 50 years. Smith's plus/minus was as good as it was because he was playing with Bobby Orr. Smith was still a quality defenceman; he was the defence partner of the greatest defenceman of all time, after all. But his numbers were inflated by playing for what was the most dominant offensive team in NHL history. Indeed, as I mentioned already, all plus/minus leaders have always come from strong teams. This is a major problem for SAT as well; indeed, this is why most people quote the team-relative version (SAT rel) instead of raw SAT numbers. We will get back to this when we talk about SAT in more detail.

3. COMPETITION EFFECTS

If you play against stronger players, you will get outscored more severely than if you play against weaker players; this is fairly uncontroversial. This has rarely been a big problem with plus/minus because in the NHL good players tend to play against good players, not the opposite. However, this does mean that the plus/minus of good players may be lower than it would be if it were "unbiased." This problem exists also for SAT; indeed, it has often been pointed out as a major factor influencing SAT. We will get back to this later as well.

4. ZONE START EFFECTS

Players who start all their shifts in their own zone are at a disadvantage against those who are always deployed in an offensive role. If you start a 50-second shift with a faceoff in your own zone, your team will on average be outscored by about three to two. Luckily, there are no NHL players who start all their shifts in the defensive zone, although Manny Malhotra often comes close. In 2014–15, Nashville's Paul Gaustad started 506 times in his own zone and only 64 times in the offensive zone. Seen in this light, his +7 mark is downright spectacular. Zone start effects also exist for SAT, and they are even more pronounced for SAT than for plus/minus.

5. GOALTENDING

Sergei Gonchar makes a defensive mistake, leading to a scoring opportunity—only for Carey Price to make a spectacular save. This defensive gaffe won't appear anywhere on Gonchar's plus/minus, thanks to Price's heroics. In general, players who play in front of above-average goaltenders will see their plus/minus artificially inflated, while those who protect mediocre goaltenders will see their numbers sink. This is one of the great benefits of shot-based metrics: they are completely unaffected by the goaltending at either end of the rink, and thus they can focus on what the skaters are actually doing. However, even this will have some unexpected consequences, as I will discuss later.

6. SAMPLE SIZE

This is what has ultimately killed plus/minus, so much so that I will give it its own section.

The Fatal Problem of Sample Size

Let us take a fairly elite, theoretical NHL player. Let's ignore his teammates, his opponents, and his zone starts for now. To make sure we don't confuse him with any actual NHL players, let's call him Davel Patsyuk. Davel is so good that, while he is on the ice, his team takes 34 shots per 60 minutes and allows only 29; this means that they get 54% of all the shots taken while he is on the ice. This is pretty good; there were only two teams in the NHL in 2014–15 that averaged more than 53.1% of the shots (the Los Angeles Kings and the New York Islanders, for you people keeping score at home).

Let's also assume Davel has no impact on shooting percentage. The average NHL team scores on about 8% of its shots at even strength. This means that, overall, his team will score 2.72 goals per 60 minutes while he is on the ice (34 shots times 8%) and allow 2.32 goals (29 shots times 8%). If the team could have Davel on the ice all season and play all 82 games at even strength, then they would on average score 223 goals and allow 190, giving them about 103 points. Davel is really good.

So the first question is, in practice, what do we expect Davel's plus/minus to be?

On average, we saw that his team outscores the opponents by 0.4 goals per 60 minutes. If he plays a total of 1,260 minutes at even strength all season, a fairly robust total for a forward, his plus/minus will be, on average, 0.4 times 1,260 divided by 60, which equals 8.4.

Wait, that's it?

Yup. This elite player, whose characteristics would make him one of the top 30 or so forwards in the league, would have an average plus/minus of 8.4 (on a team with average teammates). The second question is, figuring in puck luck, what is the range of plus/minus stats he can expect to get over the course of an entire season?

I won't keep you in suspense.

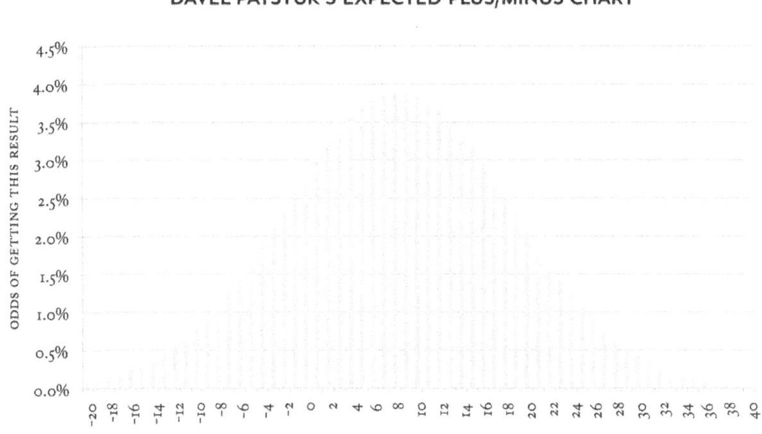

That's right. Even over a single season, this elite player, whose expected plus/minus is 8.4, has a 22% chance of having a plus/minus that is zero or negative and has a 5% chance of posting −9 or below. Conversely, one season out of five he will score +18 or better, and one season out of 20 he will score +26 or better. All of this is assuming that he remains at the exact same talent level year after year, with perfectly average teammates and perfectly average goaltending. Add some variation to one of those variables and the numbers get even noisier.*

[* I arrived at these numbers by modelling all the goals scored during the season as a Poisson distribution. With the quoted numbers, the average number of goals for is 60 and the average number of goals against is 50.4. Therefore, the total number of goals is 110.4, which is λ in the Poisson distribution, and the standard deviation is the square root of this (110.4), which is 10.5.]

Ultimately, this is why plus/minus, despite its theoretical "perfection," fails. There is simply too much noise to make an intelligent analysis of a player over any reasonable time frame. Plus/minus, adjusted for team strength, is extremely good for judging players, but only over the course of several years. As an example, here is the list of the best players in adjusted plus/minus from 2005-06 to 2014-15, a span of nine seasons.

ADJUSTED PLUS/MINUS LEADERS, 2005-06 THROUGH 2014-15*

NAME	TEAM	GP	+/-	+/- ADJUSTED
Pavel Datsyuk	DET	615	+208	182.4
Sidney Crosby	PIT	550	+124	178.3
Joe Thornton	BOS/SJS	698	+153	174.6
Marian Hossa	ATL/PIT/DET/CHI	623	+147	162.2
Henrik Sedin	VAN	692	+164	155.2
Chris Kunitz	ANA/PIT	638	+151	146.0
Daniel Sedin	VAN	664	+148	143.1
Martin St. Louis	TBL/NYR	696	-13	132.7
Daniel Alfredsson	OTT/DET	617	+101	123.7
Nicklas Lidstrom	DET	548	+173	122.2
Alexander Semin	WSH/CAR	526	+82	116.4
Jonathan Toews	CHI	484	+141	115.7
Ryan Getzlaf	ANA	633	+109	115.0
Lubomir Visnovsky	LAK/EDM/ANA/NYI	562	+22	113.3
Henrik Zetterberg	DET	619	+144	111.7
Alexander Ovechkin	WSH	679	+47	107.4
Rick Nash	CBJ/NYR	629	+17	105.9
Marian Gaborik	MIN/NYR/CBJ/LAK	515	+80	98.1
Alexandre Burrows	VAN	618	+122	97.1
Zach Parise	NJD/MIN	617	+69	94.1

[* Raw data for plus/minus calculations obtained from the NHL's website, accessed May 9, 2015, http://www.nhl.com.]

While you may argue with the exact order of players on this list, there is little doubt that this is an extremely convincing list of players who helped their teams outscore the opposition at even strength. There may be a few passengers (it's uncertain how well Kunitz, Burrows, or Semin would have done with weaker linemates), but these are still very strong players. And while I don't have exact data for it, we can still see the bias of zone starts: I wouldn't be surprised if the Sedins dropped quite a bit down the list if we took zone starts into account.

You will also notice that the 20th player on our list, Zach Parise, has an adjusted plus/minus of 12.5 per 82 games (94.1 divided by 617 games). That means that, as per our previous example, not many players can sustain a "true" plus/minus of +10 or above year after year.

What If We Used Shots?

However, in the real world we are often trying to analyze players over a shorter time frame. Given that sample size is an issue with plus/minus, what if, instead of using goals, we tried using shots to analyze a player's calibre? We know that, on average, Davel's team will outscore the opposition by five shots per 60 minutes while he is on the ice. Again taking puck luck into account, what is the distribution of *those* results at the end of the season?

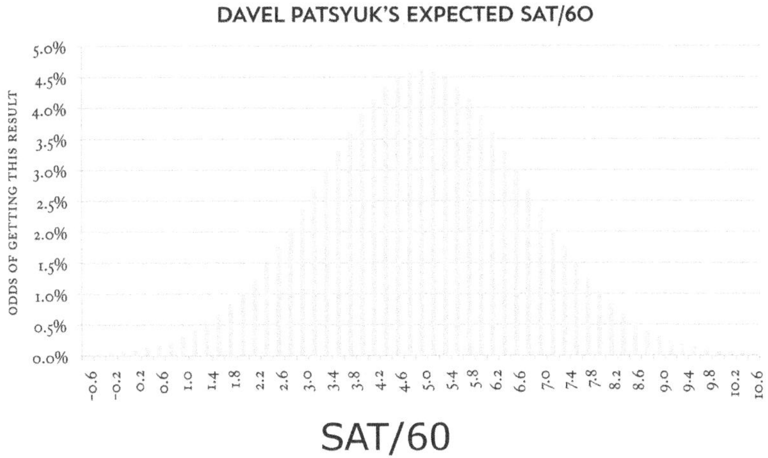

Clearly, the range of results is much narrower in this case. Obviously the average SAT/60 is +5, but the vast majority (93%) of seasons it will fall between +2 and +8, a fair representation of his talent level, and only 0.2% of seasons will produce a result of 0 or worse.

Fundamentally, this is the reason to use shot-based metrics. The sample size of shots is 12 times that of goals, and the sample size for unblocked shot attempts, including misses, is about 16 times that of goals. This means that shot-based metrics can give us an idea of a player's performance that is impossible to get from goal-based metrics.

However, shots have been tracked by the NHL for almost 50 years. Why did shot-based metrics emerge so recently, and how did they take over statistical analysis of the NHL?

The Unstoppable Rise of Corsi

On December 26th, 2010, the Dallas Stars were fifth in the NHL, with 46 points in 35 games played, while the New Jersey Devils were dead last, with 20 points in 34 games. On December 28th, a relatively unknown blogger named Hawerchuk (whose identity will be revealed in a moment) implied that the Devils might actually be the better team and asked on his website, "Which team will have more points over the rest of the season?" Here are the answers: 45% of respondents answered that Dallas would have 10 or more points than New Jersey, 19% answered that New Jersey would have 10 or more points than Dallas, and 36% answered in between.*

[* Gabriel Desjardins (Hawerchuk), "Poll of the Day: New Jersey vs Dallas," *Arctic Ice Hockey* (blog), December 28, 2010, http://www.arcticicehockey.com/2010/12/28/1896389/poll-of-the-day-new-jersey-vs-dallas.]

Dallas went on to obtain only 49 points in their remaining 47 games, narrowly missing the playoffs, while the Devils piled on 61 points in their last 48 games. The next season, those same Devils would hit 102 points in the standings and get all the way to the Stanley Cup Final, ultimately losing to the Los Angeles Kings in six games.

How did Hawerchuk know?

Hawerchuk, also known as Gabriel Desjardins, had participated over the previous couple of years in the emergence of shot-based metrics as a new means of measuring the performance of teams and players. Between 2007 and 2010, a small group of passionate fans, including Tyler Dellow, Vic Ferrari (real name Timothy Barnes), Sunny Mehta, JLikens (real name Tore Purdy), and Desjardins, had discovered the benefits of shot-based metrics and had developed a huge appreciation for the shot-based metric they called Corsi.

What Is Corsi?

The Corsi stat, which has since been renamed shot attempts (SAT) by the NHL, is the sum of all shots directed at the opposing net while a player is on the ice minus all shots directed at his own net. In simple terms, the formula is:

Corsi = shot attempts for − shot attempts against

Shot attempts is thereby the sum of all shots, missed shots, and blocked shots taken by the team. In this respect, Corsi is like plus/minus except for shot attempts—above zero is good, and below zero is bad. Corsi can also be expressed on a per-60-minute basis, or as a percentage:

Corsi percentage = shot attempts for/(shot attempts for + shot attempts against)

In this case, 50% is average and the higher the better.

Corsi was named for Jim Corsi, the Buffalo Sabres' goaltending coach. Ferrari named the new statistic after Corsi because he had heard the Buffalo Sabres' general manager Darcy Regier talking about shot attempts and wanted to name the statistic after him, but he liked the sound of "Corsi" better—not to mention his moustache.* Coincidentally, it turns out that Jim Corsi had indeed pioneered the notion of using all shot attempts to measure a goaltender's workload. However, he and the Sabres had never used it as a measure of shot differential, and they certainly didn't name it.

[* Technically, Corsi wasn't new, as early versions were used by Lloyd Percival and Anatoli Tarasov half a century ago and by hockey analysts like Tom Tango prior to Ferrari. However, Ferrari and company are generally credited with doing all the math, establishing its persistence and the link with possession, and popularizing its usage in its current, modern form; see also Bob McKenzie, "The Real Story of How Corsi Got Its Name," TSN, September 27, 2014, accessed August 31, 2015, http://www.tsn.ca/mckenziethe-real-story-of-how-corsi-got-its-name-1.100011.]

The Link to Puck Possession

However, these young pioneers had discovered a new use for this statistic: to estimate puck possession.

Before we continue, it is important to remember that all discoveries are, to some extent, products of their times. In 16th-century Europe, the new technology of the printing press allowed the massive printing and distribution of books, spreading knowledge and piquing curiosity all over the continent, leading to the Renaissance. The early 20th century saw massive (and tragic) advances in military technology, spurred by international rivalries and world wars. In similar fashion, the development of possession metrics was a discovery waiting to happen in the NHL of the late 2000s, because the NHL of that era had two things that previous eras had lacked: accurate shot-tracking technology and the Detroit Red Wings.

While it may seem quaint to think of it now, especially since they won "only" one Stanley Cup over those years, the Red Wings were without a doubt *the* dominant team of the post-lockout era, especially between 2005–06 and 2008–09. More than any other team, the Red Wings had adapted to the famous "crackdown on obstruction" that the NHL, hoping to stimulate more offence, implemented after the 2005 lockout. Aided by the charity point, the 2005–06 Red Wings accumulated 124 points, the second-highest total of the modern era (behind only the 1995–96 Red Wings).

Yet the Red Wings achieved all of this even though, by traditional metrics, their players were not perceived as being among the elite of the league.

- Chris Osgood has always been considered an above-average goaltender who made it onto an elite team.
- Pavel Datsyuk, despite an otherworldly season in 2007–08, finished ninth in Hart Trophy voting that year, behind Alexei Kovalev!
- Henrik Zetterberg was respected but not revered.
- The only exception to this lack of recognition was Nicklas Lidstrom, who by then had been around the league long enough that it was clear he was a god simply masquerading as a human.

While the league was abuzz with the new, young superstars Sidney Crosby, Alexander Ovechkin, and Evgeni Malkin, the Red Wings were dominating all other teams but without the individual stats to back it up. It is against this background that the statistics for estimating possession were established.

Two things were clear about the Red Wings: they *always* had the puck, and they were very consistent. Take a look at the shot differential by team in 2007–08:

SHOT DIFFERENTIAL BY TEAM, 2007–08*

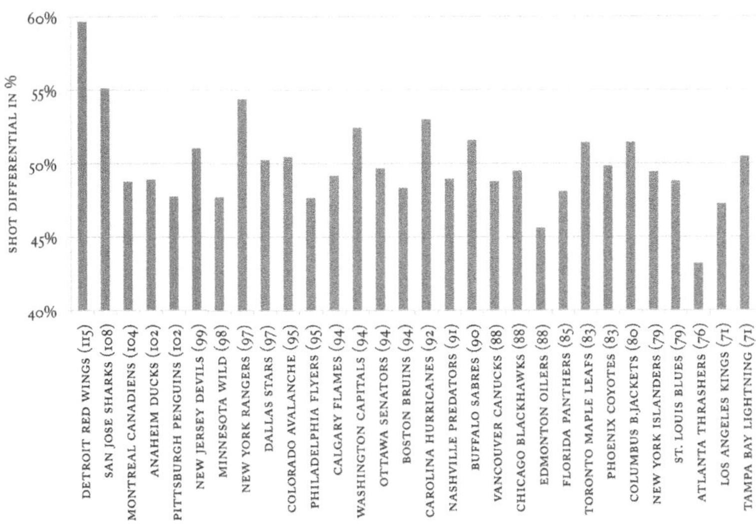

TEAMS (POINTS IN PARENTHESES)

[* Raw shot data obtained from the NHL's website, accessed August 31, 2015, http://nhl.com.]

No, that's not a typo. The Red Wings got 59.7% of the shots, outshooting their opponents by 1,173 shots, light-years ahead of their nearest competitors, the San Jose Sharks. More importantly, the top two teams in shot differential were also the top two teams in the standings, even though the relationship wasn't as strong in the middle of the league. There was clearly something going on here.

Looking into shot differential more deeply, the Corsi fanatics discovered something that was very interesting: shot differential was a skill that was *persistent.* This comes back to our earlier observations about sample size. When a team has taken 55% of the shots over 20 games, that meant they had outshot them about 660–540; a difference of 120 shots wasn't likely to be due to luck. Conversely, when a team got 55% of the goals over 20 games, that meant they had outscored the opposition roughly 55–45. Five lucky bounces and half that difference would disappear. Five lucky bounces aren't that many over 20 games.

Predicting Future Success

Let me show the value of Corsi in a different way. I will give you the final NHL standings of 2013–14 and ask you to predict, based on only this information, what the final standings of 2014–15 were. You have no information on player movement, injuries, or anything else. What would be your best guess?

FINAL NHL STANDINGS, 2013–14*

TEAM	GP	PTS	TEAM	GP	PTS
Boston Bruins	82	117	Dallas Stars	82	91
Anaheim Ducks	82	116	Washington Capitals	82	90
Colorado Avalanche	82	112	Arizona Coyotes	82	89
San Jose Sharks	82	111	Nashville Predators	82	88
St. Louis Blues	82	111	New Jersey Devils	82	88
Pittsburgh Penguins	82	109	Ottawa Senators	82	88
Chicago Blackhawks	82	107	Toronto Maple Leafs	82	84
Tampa Bay Lightning	82	101	Winnipeg Jets	82	84
Los Angeles Kings	82	100	Carolina Hurricanes	82	83
Montreal Canadiens	82	100	Vancouver Canucks	82	83
Minnesota Wild	82	98	New York Islanders	82	79
New York Rangers	82	96	Calgary Flames	82	77
Philadelphia Flyers	82	94	Edmonton Oilers	82	67
Columbus Blue Jackets	82	93	Florida Panthers	82	66
Detroit Red Wings	82	93	Buffalo Sabres	82	52

The simplest guess would be, of course, "the same standings." That would be a reasonable assumption if we thought that the 2013–14 standings were a perfect representation of each team's ability and that the teams didn't change from year to year. In practice, of course, both of these things aren't true. We know the second, but it's the first one that's important here: even over the course of 82 games, points in the standings aren't a very accurate representation of a team's true talent. This is because, ultimately, there is a lot of luck involved in hockey. Obviously not 100%, otherwise we wouldn't be playing hockey, we would be playing Snakes and Ladders. But neither is hockey a game of pure ability, such as chess, with no luck whatsoever.

Historically, we will find that standings points will predict only 24.5% of a team's success the next season, leaving 75.5% unexplained.[29] But if points in the standings aren't the best we can do, then what is?

---

29 The numbers I am quoting here are known as R^2 (R squared), which quantifies how much of a variable is explained by another variable. This is related to the correlation coefficient, R, but R^2 values are smaller. If R = 0.6, the R^2 will be 0.6x0.6 = 0.36. Just don't get them confused! Remember that when someone says that she's found a "strong" correlation of 0.2, that means her R^2 is 4%, so 96% of the variation is still unexplained! Rob writes about correlation coefficients in a little more detail in the team-building chapter.

The next-best answer comes to us from the ultimate pioneer of sports statistics, Bill James. James realized that by using runs scored and runs allowed, he could better predict how many games a team would win in the future than by simply using wins and losses. He developed what he called "Pythagorean expectation," whose name comes from the fact that the runs scored and runs allowed are squared in his equation, similar to the Pythagorean theorem for calculating the hypotenuse of a triangle.* James's formula was as follows, where RS is runs scored, RA is runs allowed, and ^2 is value squared.

[* "Pythagorean Expectation," Wikipedia, accessed August 31, 2015, https://en.wikipedia.org/wiki/Pythagorean_expectation.]

win% = RS^2/(RS^2 + RA^2)

James developed this formula over 35 years ago and there have been many refinements over the years, but the core of its logic has been well established, even for hockey. Goals for and goals against are better predictors of future wins than wins themselves. The reason for this is fairly straightforward, even though it goes against what all hockey commentators and insiders have tried to drill into us for decades: winning a blowout is a far more convincing display of skill than winning a close game, and indeed winning a close game more often than not comes down to chance. When we use goal differential instead of points, we find that we can predict 29% of a team's success the next season.

Can we do better? Obviously, since I'm asking, the answer is yes. What if, on top of the team's goal differential, I also gave you their shot differential. Would this help? The answer depends on the era of hockey we are looking at, but in the modern era the answer is a strong yes. Using just shot differential, we can predict 27% of a team's success the next year, making it an even better predictor than points, albeit a slightly worse one than goal differential. But if we combine goal differential and shot differential, we find that we can predict 35.7% of each team's success the next year. This is a much more accurate predictor than goal differential by itself. Shot differential is a clear skill above and beyond what we can measure from goals.

The reason why 35.7% is smaller than the sum of 29% and 27% is because goal differential and shot differential are highly correlated; teams that outshoot their opponents tend to outscore them, all other things being equal. In fact, goal differential is simply the sum of two numbers: goal differential due to shots plus goal differential due to percentages. Which is the dominant skill? If we try to predict a team's success based solely on shooting percentage, we find that we are able to predict only 7% of the next season's success. So the major driving factor behind goal differential seems to be shot differential. But as we saw earlier with Davel Patsyuk, shot differential is much easier to measure accurately than goal differential, especially over a short period of time.

This, fundamentally, is the major argument for using shot-based metrics. We all agree that, at the end of the day, what we care about is wins. Goals are the major driver of wins, but shots are the major driver of goals, and shots occur much more frequently than goals, making them easier to count with a small sample. Thus, counter-intuitively, over the short term current shots are a more accurate predictor of future goals than current goals. Over the medium term (a full season), a blend of goals and shots is the most accurate predictor of future goals. And in the long run (several seasons), goals alone would be the most accurate predictor.

But in the long run, we are all dead.[30] Most often, general managers, coaches, commentators, and fans try to make assessments and decisions over samples of a few games or fractions of a season. In this context, Corsi becomes the most accurate prediction tool available.

---

30 Quote from famed economist John Maynard Keynes. John Maynard Keynes, A Tract on Monetary Reform (London: Macmillan, 1923).

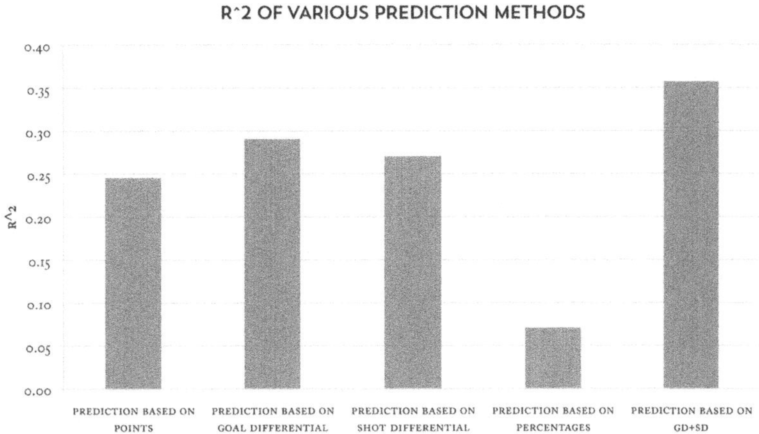

Riding the Percentages

The second, much more controversial corollary to Corsi was this: while shots and puck possession are repeatable skills, shooting and save percentages are not. More than anything else, this statement is and always has been the source of the greatest skepticism regarding possession metrics.

Shooting Percentage

The argument that shooting percentage isn't a repeatable skill can be demonstrated mathematically, as in my example about sample size, but it can also be observed. In 2013–14, among players with 150 shots or more, the top-five players in shooting percentage were Gustav Nyquist, Joe Pavelski, Paul Stastny, David Backes, and Chris Kunitz. A fine selection of forwards, to be sure, and indeed I would feel good about my Cup chances if all were among my top six. How did they do in 2014–15? All of them regressed both in shooting percentage (Sh%) and goals per game (G/GP).

2013–14 SHOOTING PERCENTAGE LEADERS, BEFORE AND AFTER*

NAME	TEAM	SH% PRIOR TO 2013-14	2013-14 G/GP	2013-14 SH%	2014-15 G/GP	2014-15 SH%
Gustav Nyquist	DET	N/A	0.49	18.3%	0.33	13.8%
Joe Pavelski	SJS	10.0%	0.50	18.2%	0.45	14.2%
Paul Stastny	COL/STL	12.3%	0.35	16.7%	0.22	11.2%
David Backes	STL	11.6%	0.36	16.4%	0.33	14.2%
Chris Kunitz	PIT	12.8%	0.45	16.1%	0.23	10.0%

[* Shooting percentage data obtained from the NHL's website, accessed August 31, 2015, http://www.nhl.com.]

Their average shooting percentage dropped from 17.1% to 12.7%, a drop of over a quarter, and their average goals-per-game dropped from 0.43 to 0.31, a decrease of 28%. The reason why is fairly obvious: in the NHL, with the exception of a handful of superstars like Steven Stamkos, nobody actually has the skill to score on 17% of their shots. When they do, it is almost always due to luck. Fans always want to believe that a player has found "a new level" to his game, and while it is true that in their early years players do improve rapidly, it is almost unheard of for a player to improve his shooting skill dramatically at age 26 or older. And when a young player like Nyquist does emerge, while he is probably good, it is not statistically likely that he is the next Stamkos.

That's not to say that players don't have some control over their shooting percentage. Some, like Wayne Simmonds, have an abnormally high shooting percentage due to their style of play—specifically, parking themselves in the slot and scoring off deflections. Others, like Brad Richards, are the opposite: Richards plays the point on the power play, so his shots, when they do make it to the net, are fairly weak and easily stopped. However, if Simmonds were to score on 19% of his shots one season (instead of the 13% he has averaged over his career), we would not expect him to maintain it the following year.

What's true for players is doubly true for teams: barring a massive change of personnel, teams that exhibit a large rise or fall in their shooting percentage are not likely to maintain it over the long term. This, in a nutshell, was what convinced Desjardins, in our opening example, that Dallas's success was not likely to be maintained. To see your shooting percentage rise after adding Corey Perry to your team is one thing, but to see it jump for no apparent reason is not likely to signal a permanent increase in skill level.

Still don't believe me? Here are the six teams whose even-strength shooting percentage jumped the most between 2012–13 and 2013–14 and what happened to them the subsequent year.

THE FATE OF TEAMS WITH LEAPS IN SHOOTING PERCENTAGE, 2014–15*

TEAM	2012-13	2013-14	LEAP!	2014-15	FATE
Colorado Avalanche	7.1%	8.8%	+1.7%	8.7%	-0.1%
Ottawa Senators	5.9%	7.5%	+1.6%	8.1%	+0.7%
Florida Panthers	5.8%	7.3%	+1.6%	7.3%	0.0%
Detroit Red Wings	6.7%	8.2%	+1.4%	7.3%	-0.8%
Anaheim Ducks	8.6%	9.8%	+1.1%	8.3%	-1.5%
Boston Bruins	7.4%	8.5%	+1.1%	7.1%	-1.4%

[* Shooting percentage data obtained from the NHL's website, accessed August 31, 2015, http://www.nhl.com.]

Of these six teams, one of them, the Ottawa Senators, managed to increase their shooting percentage again the next year, two treaded water, and three of them fell back, including two, Anaheim and Boston, who fell *hard.* Also, the two teams who actually managed to maintain their shooting percentage two years in a row, Ottawa and Florida, had the two worst shooting percentages in the entire league in 2012–13, indicating that it was probably that lockout-shortened season that was the anomaly.

What about teams whose shooting percentage drops? Is that more likely to be "real"?

THE FATE OF TEAMS WITH DROPS IN SHOOTING PERCENTAGE, 2014–15*

TEAM	2012-13	2013-14	DROP!	2014-15	FATE
Los Angeles Kings	7.6%	6.6%	-1.0%	7.6%	1.0%
Dallas Stars	9.5%	8.2%	-1.2%	8.8%	0.5%
Montreal Canadiens	8.9%	7.5%	-1.3%	7.9%	0.4%
Pittsburgh Penguins	9.9%	8.3%	-1.6%	7.5%	-0.8%
Toronto Maple Leafs	10.7%	8.5%	-2.2%	7.4%	-1.1%
Tampa Bay Lightning	10.0%	7.8%	-2.2%	9.1%	1.3%
Buffalo Sabres	8.0%	5.8%	-2.2%	7.2%	1.4%

[* Shooting percentage data obtained from the NHL's website, accessed August 31, 2015, http://www.nhl.com.]

Of the seven teams whose shooting percentage dropped by more than 1% between 2012–13 and 2013–14, five of them bounced back the following year. Again, Toronto had the highest shooting percentage in the entire league in 2012–13, so this was likely to be unsustainable to begin with, although I doubt anyone thought they'd be so bad just two years later. Obviously, as I said earlier, personnel changes *can* have an effect. The Lightning of 2014–15 are almost nothing like the Lightning of 2012–13: Tyler Johnson and Nikita Kucherov weren't even in the league, while the 2012–13 team had Martin St. Louis, Vincent Lecavalier, and Teddy Purcell among its top four forwards. But when a team retains the bulk of its players year to year, especially its top players, and sees a huge rise or drop in its shooting percentage, you should smell a rat.

So is shooting percentage a myth? The greatest whipping boy of the early years of advanced statistics was Ilya Kovalchuk. Kovalchuk was an easy target: his team, the Thrashers, was, to put it mildly, terrible, which ensured that he was a massive minus in both plus/minus and Corsi every year of his career in Atlanta. His greatest skill, the power play, is not measured by Corsi, which is fundamentally an even-strength measurement. And he was one of the few NHLers who was able to put the puck in the net at an above-average rate, a skill that was explicitly ignored by Corsi. Needless to say, Corsi made Kovalchuk look very bad, and he was often panned by the analytics community as the prototypical example of an overpaid superstar, whose headline numbers, in particular goals scored, made everyone forget his flaws.

Now that he has left the NHL, remembered mostly for a long-term contract that turned messy, it's easy to forget that, within two years of being acquired, Kovalchuk helped the New Jersey Devils reach the Stanley Cup Final. Kovalchuk was always very talented, but his talents were not well captured by Corsi. Kovalchuk had the best shot in the NHL at the time; only Ovechkin was in the same league. He had a career shooting percentage of 14.1%, despite often taking shots from 30 to 50 feet from the net. Shot volume doesn't matter quite as much when you have Kovalchuk taking the shot.

Possession stats are explicitly designed to ignore shooting percentage. However, let me be clear: shooting percentage skill *does exist* and is indeed a significant contributor to many teams' and players' success. In the 2014 edition of *Hockey Abstract,* I showed that first-line players outscore their opponents by an average of 0.39 goals per 60 minutes and that a little over half of that was due to having a better shooting percentage than their opponents.[31] The problem with shooting percentage skill isn't that it doesn't exist, it's that it's not precisely measurable over a period of time shorter than a season or two. When we see a player on a scoring streak, we *want* to believe that the player has cracked the code, that he's found another level to his game, but it's just an illusion. Within two weeks, he'll go back to his regular level of scoring, and fans and the media will be left to wonder where the magic went.

Save Percentage

Save percentage is the mirror image of shooting percentage, with one large difference: while teams are composed of multiple skaters, and the team shooting percentage is the sum of the shooting of all skaters, teams employ a small number of goaltenders, of whom one often plays more than 50% of the games in a single season. As such, save percentage behaves as a hybrid between a team statistic and an individual statistic.

---

31 Tom Awad, "What Makes Good Players Good? Part 1: Forwards," in Hockey Abstract 2014, by Rob Vollman (author, 2014), 53.

It used to be that there was a massive gulf between the best goaltenders in the NHL and the worst. During the 1990s, Patrick Roy, Ed Belfour, and Dominik Hasek were much better than everyone else; the three of them won 10 of the 11 Vezina Trophies between 1989 and 1999. At the other end, as the NHL expanded from 21 to 30 teams, there was a demand for more goaltending talent, and the talent pool diluted. The standard deviation of career save percentage for all goaltenders who played 100 or more games between 1989 and 1999 was 11.1 points, or 0.011 on a traditional three-digit save percentage (e.g., 0.900).

Over the last 10 years, things have stabilized. While there are still elite goaltenders like Henrik Lundqvist, there is much more talent. The popularity of Roy, and later Martin Brodeur, led to an influx of goaltenders from Quebec, while the Nordic countries also started producing large numbers of quality netminders. The last eight Vezina Trophies have been won by seven different players. The standard deviation of save percentage for all goaltenders who played 100 or more games between 2005 and 2015 was just 7.5 points. When we consider that the standard deviation expected over this period due just to luck is around 3.4 points, we can see that the variation in goaltending ability has reduced significantly.[32] By 2010, analyst Sunny

32 This is a rough approximation. These 78 goaltenders averaged 691 goals against on 7,799 shots against. The expected error on 691 goals against is about 26.3 goals, and 26.3 divided by 7799 is 0.0034.

Mehta (now of the New Jersey Devils) even claimed: "No one has conclusively shown a meaningful difference in skill between NHL goaltenders."* I would not go that far, but my analysis in 2010 indicated that, over a single season, 66% of a goaltender's save percentage is due to luck.** But Rob covered this area in plenty of detail in his best puck stopper chapter, so I'll leave it at that.

[* Sunny Mehta, "New Jersey Devils vs Philadelphia Flyers Preview," *Arctic Ice Hockey* (blog), April 13, 2010, http://www.arcticicehockey.com/2010/4/13/1419530/new-jersey-devils-vs-philadelphia.]

[** Tom Awad, "Numbers on Ice: How Much Skill Is There in Goaltending?," *Hockey Prospectus* (blog), May 7, 2010, http://www.hockeyprospectus.com/puck/article.php?articleid=558.]

Since teams tend to keep the same number-one goaltender from year to year, that means that some portion of the team's save percentage should be a skill that will remain constant from year to year. In practice, that is what we observe: the correlation between a team's even-strength save percentage this year and next year is 0.3, which is not very high but certainly indicates some level of persistence, but in practice, it is hard to count on.[33] Here are the top

---

33 All save percentages in this section are even strength and score adjusted to make them more representative of true skill, but you would observe the same pattern using unadjusted even-strength numbers. There is more analysis

five teams in even-strength save percentage in 2013–14 and how they fared in 2014–15.

TOP FIVE TEAMS IN EVEN-STRENGTH SAVE PERCENTAGE, 2013–14 AND 2014–15*

TEAM	2013-14 SV%	2014-15 SV%
Boston Bruins	93.9%	92.9%
Los Angeles Kings	93.5%	92.6%
Minnesota Wild	93.2%	91.7%
New York Rangers	93.0%	93.0%
Phoenix Coyotes	93.0%	91.6%
Average	93.3%	92.4%

[* Save percentage data obtained from the NHL's website, accessed August 31, 2015, http://www.nhl.com.]

All of them except New York regressed, despite the fact that they all, except Minnesota, kept the same number one from one year to the next (and Minnesota stumbled onto Devan Dubnyk, who was even better than Josh Harding had been the previous year). How did the teams at the bottom fare?

BOTTOM FIVE TEAMS IN EVEN-STRENGTH SAVE PERCENTAGE, 2013–14 AND 2014–15*

---

of such details in Rob's chapter on the league's best puck stopper (see pages 169–227).

TEAM	2013-14 SV%	2014-15 SV%
Chicago Blackhawks	91.4%	93.5%
New York Islanders	91.2%	91.4%
Florida Panthers	91.0%	92.3%
Nashville Predators	91.0%	93.0%
Calgary Flames	90.9%	92.3%
Average	91.1%	92.5%

[* Final standings data obtained from the NHL's website, accessed August 31, 2015, http://www.nhl.com.]

Again, we see the same thing. All five teams improved from their league-trailing performance, and Chicago and Nashville were actually in the top five in 2014–15. Incredibly, these five teams did better, on average, in 2014–15 than the top five teams did. While this is unusual, the lack of correlation between one season's success and the next is not.

The bottom line is that, over a single season, save percentage numbers, both for individual goaltenders and teams, are quite noisy and unreliable. This is another reason why individual plus/minus numbers and even team goal-differential numbers are less accurate than possession metrics for evaluating skill.

PDO and Regression to the Mean

Among the most obscurely named statistics to emerge from the hockey analytics revolution is a statistic called PDO, named after one of its inventors, Brian King, a blogger who went by that nickname. The formula for PDO is extremely simple:

PDO = on-ice shooting percentage + on-ice save percentage

The end result is expressed out of 1,000. So if your team's shooting percentage while you are on the ice is 9.5% and its save percentage is 91%, then your PDO will be 1,005 (95+910). PDO can be calculated for teams as well as for individual players. PDO has been rebaptised by the NHL as SPSV%, a name that at least has the benefit of explaining exactly what it is, shooting plus save percentage.

DISTRIBUTION OF PDO, ONE SEASON VS THREE SEASONS*

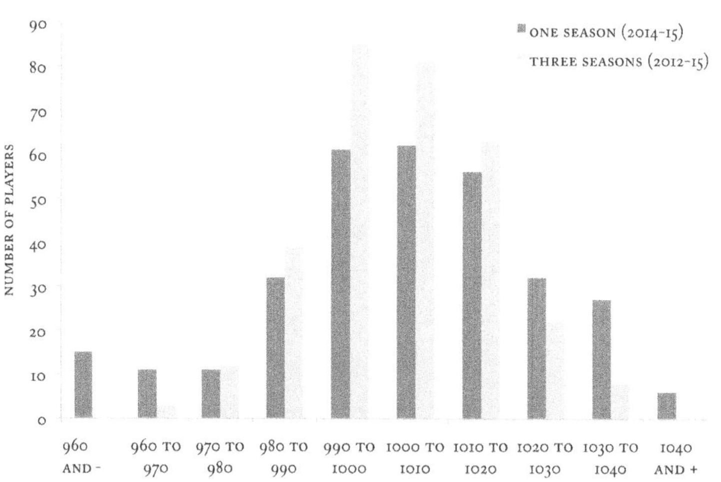

[* Raw PDO data obtained from the NHL's website, accessed August 31, 2015, http://nhl.com.]

Desjardins once called PDO "the single most useful statistic in hockey,"* and while I wouldn't go that far, understanding it gets you a long way to understanding why possession stats work. SPSV% doesn't actually represent one specific thing, since adding shooting percentage and save percentage is, by itself, fairly meaningless. SPSV% represents the contribution that percentages made to your on-ice goal differential. When a player's (or team's) SPSV% is significantly above 1,000, luck is going his way—he's getting the bounces. Conversely, when SPSV% is well below 1,000, the bounces are going against him. This is a statistic that tends to regress to the mean, where the mean is 1,000. SPSV% is thus an easy way to tell if a player's or team's successes are sustainable or if the tide is likely to turn against them sooner or later. We can see this by comparing the single-season distribution of SPSV% to the three-year distribution. Over a single season, many players obtain a result of 1,030 and above or 970 and below, but over three seasons these outliers disappear almost completely.

[* Gabriel Desjardins, "PDO: If You Were Going to Understand Just One NHL Statistic," *Arctic Ice Hockey* (blog), October 28, 2011, http://www.arcticicehockey.com/2011/10/28/2520115/pdo-if-you-weregoing-to-understand-just-one-nhl-statistic.]

Remember that while we may measure a player's on-ice SPSV%, it is not really an individual statistic, since it is affected by the calibre of the player's teammates and the goaltending behind him. However, players do have some impact on their on-ice shooting percentage, and consequently their long-term SPSV% can trend slightly above or slightly below 1,000.

Analytics wonks like to use the expression "regression to the mean," which is useful but misleading. When we say that a player or team will regress to the mean, what we actually mean to say is that he will regress to his true talent level. If Steven Stamkos has scored seven goals on 15 shots in the last five games, I don't expect him to be able to keep it up. Similarly, if Scott Gomez hasn't scored a goal all season, I expect that he will at some point (although it could still take a while). This doesn't mean that I think they will both regress to the same shooting percentage. Gomez probably has a "true" shooting percentage of 5% or 6%, while Stamkos's true talent level is about 17%. In the long run, both players should regress to their own level.

As an example of how to use SPSV%, let us look at the 2013–14 statistics. David Krejci led the NHL with a +39 plus/minus rating and had an impressive 69 points, helping the Boston Bruins finish first overall in the league. At the other end, Alexander Edler mirrored Krejci's performance, finishing with a −39 as his Vancouver Canucks missed the playoffs by a fair margin. Had either player earned this number, or was it mostly a matter of luck?

We can see that Krejci's USAT% of 51.8% was good but not world dominating. The real reason his plus/minus was so good was because his Bruins scored on 9.6% of their shots while he was playing but got scored on only 5.2% of the time, thanks mostly to the superlative goaltending of Tuukka Rask. Krejci's SPSV% was thus 1,044, one of the highest in the league.

Impressively, Edler's USAT% was even better than Krejci's, at 51.9%. Of course, as a defenseman, Edler had little control over the fact that his teammates scored on a miserable 3.6% of their shots while he was on the ice. While Vancouver's save percentage with him was a respectable 91.6%, that still gave him a SPSV% of 952, a mark that is unsustainable in its mediocrity.

What did we see in 2014–15? Krejci's USAT% dipped down to 48.6%, but his SPSV% remained high, at 1,021, mostly because he was still playing in front of superior goaltending in Boston. But 1,021 is much more "normal" than 1,044, and he returned to a plus/minus of just +7. Meanwhile, Edler's USAT% got even better, at 53.5%, but more importantly his SPSV% returned to a normal level of 997, giving him a plus/minus of +13—quite a bounce-back from −39.

This may seem like an academic exercise, since everyone knows that plus/minus is a meaningless statistic, but plus/minus at the team level is goal differential, which is the most important measure of success for a team. Like players, teams go through hot streaks and cold streaks. When the bounces are going their way, we rationalize their success: they have "built their team correctly," they "understand how to win," etc. But sometimes the puck just goes in the opposing net and stays out of yours for no good reason.

To see the effect of SPSV% at the team level, let's look at the 2013–14 standings and find the top three teams and bottom three teams in SPSV%. How do you think they fared the next season?

BEST AND WORST SPSV% TEAMS, 2013–14 AND 2014–15 RESULTS*

TEAM	2013-14 SPSV%	2013-14 POINTS	2014-15 POINTS
Boston Bruins	1,025	117	96
Anaheim Ducks	1,023	116	109
Colorado Avalanche	1,017	112	90
New York Islanders	984	79	101
Florida Panthers	982	66	91
Buffalo Sabres	981	52	54

[* Percentages and points data obtained from the NHL's website, accessed August 31, 2015, http://www.nhl.com.]

So the top three teams in SPSV% dropped from an average of 115 standings points to just 98, while the bottom three teams jumped from an average of 66 points to 82. These are HUGE jumps in each direction.

SPSV% regression is one of the most reliable phenomena in hockey statistics. This doesn't mean that *every* team will always regress back to 1,000.

Anaheim and Boston, in particular, have maintained percentages above average for years, while teams like Edmonton and Buffalo have managed the opposite feat. But whenever a team's percentages jump significantly from one season to the next, you should be wary. It's unlikely to hold.

Here is the table that explains, for teams, how highly correlated statistics are from one season to the next, where 1 represents perfect correlation (this is explained in more detail in Rob's chapter on team building).

YEAR-TO-YEAR CORRELATION SUMMARY

	USAT	SPSV%	SH%	SV%
Year-to-year correlation	0.58	0.24	0.00	0.30

We can see that USAT numbers are, by far, the most consistent on a year-to-year basis. While 0.58 may seem low, we have to remember that teams go through personnel changes, injuries, coaching changes, and other transitions that will all affect their possession numbers. In light of that, a season-to-season correlation of almost 60% is extremely stable. By contrast, shooting percentage has exactly zero correlation from one year to the next, which I found personally shocking, but the numbers don't lie. Save percentage has a year-to-year correlation of 0.30, which, as discussed in our section on goaltending, represents some skill and a large serving of luck. Finally, SPSV%, being a combination of shooting and save percentages, falls somewhere between the two.

It's important to remember that this table shows the correlation from one season to the next, so the sample size is 82 games; I specifically excluded the 2012–13 season from the calculations to ensure this. Over an even shorter period, we would expect save percentage and SPSV% to exhibit even less persistence; at 0%, shooting percentage can't possibly maintain itself any less. This is why judging teams at the quarter-point of the season, or over a six-game road trip, using goal differential numbers can be worse than useless.

Possession Numbers

The definition of Corsi includes all shots directed at the net, which includes actual shots that hit the goal, missed shots, and blocked shots. This is probably the most accurate proxy for possession, as taking any kind of shot requires your team to have had possession in the first place. But ultimately, nobody is that interested in possession. What we really want is a way to predict future goals, and in this respect it is possible that not all shots are born equal.

The easiest decision in this debate has been to drop blocked shots. As Corsi was being popularized, a blogger named Matt Fenwick proposed using only shots and missed shots as a shot differential measure, excluding blocked shots. His reasoning was that teams are capable of blocking shots or not, and since blocking a shot actually prevents it from hitting the net (and ever becoming a goal), we should exclude those shots. Sunny Mehta took this one level further, proving statistically that teams control their own rate of shot-blocking but not their opponents' shot-blocking.* The variant of Corsi that excludes blocked shots was thus baptized Fenwick, and it is now officially known by the NHL as USAT (unblocked shot attempts).

[* Sunny Mehta, "Blocked Shots: Luck or Skill?," (personal blog, currently locked), May 2010, http://vhockey.blogspot.com/2010/05/blocked-shots-luck-or-skill.html.]

The next level of analysis has been to acknowledge that not all unblocked shots are born equal, and thus USAT is imperfect as a measure of current or future goals. There would be no way to do justice to the shot quality debate that has raged over the years here, especially having covered it in comprehensive detail in the last edition,[34] but I will attempt to summarize the current body of knowledge:

- Individual shot quality is real.

There is obviously no question about this; not all shots are created equal. A rebound taken 10 feet away from the net has a five-to-ten times greater chance of scoring than a shot taken from the point.

---

34 Tom Awad, "Shot Quality," in Hockey Abstract 2014, by Rob Vollman (author, 2014), 32–43.

- In the aggregate, differences between teams in shot quality are quite small.

When I did the analysis for the 2013–14 season, I found that differences in shot quality accounted for less than 5% of team results compared to 40% for shot differential and 55% for shooting percentage differential.

- If you are attempting to predict future goals, a slight improvement on SAT or USAT is to use either weighted shots or scoring chances.

Different sources have different metrics for weighting shot quality; as of today, the gold standard for scoring chances is *War on Ice,* which has both a clear definition of scoring chances and has proven that they correlate well with future goals.*

[* Sam Ventura, "Better than Corsi: Scoring Chances More Accurately Predict Future Goals for Players," *War on Ice* (blog), January 6, 2015, http://blog.war-on-ice.com/scoring-chances-better-than-corsi/.]

- Practically speaking, the differences between all of these measures will be small.

Scoring chances are probably the best measure, but they have the smallest sample size by definition, and so they suffer from some of the problems we've discussed earlier with goals.

SAT and USAT are excellent when trying to analyze individual players, especially when performing more advanced calculations such as with or without you (WOWY).

Even pure shots, ignoring both missed and blocked shots, have two big advantages: at a team level, we have data on them for the last 48 years, and thus we have an ability to analyze them over a time period that is impossible to replicate for the others. Secondly, they are reported by every statistical source imaginable: every sports website, from ESPN to Yahoo! Sports to the NHL's site itself, has easy-to-find data for shots, in some cases going back many years, which is not necessarily true for missed and blocked shots. So if you ever find yourself wanting to do a quick analysis with whatever source of data you have at your fingertips, using shots instead of one of the more advanced measures is perfectly fine.

In the end, most of the discoveries of the early years of hockey analytics can be summarized as follows:

- Shot attempts are a good proxy for puck possession.
- Puck possession is a measurable and repeatable skill, even over the short term.
- Shooting percentage and save percentage are *not* repeatable skills in the short term, and any short-term value for them will be mostly due to luck.

These were huge insights in and of themselves, and they have immensely advanced our understanding of

NHL hockey, but they are not enough. Possession statistics are important, but individual players don't have full control over their possession numbers. Just as was the case with plus/minus, how and when a player is used will affect his numbers in huge ways. In the next section I will analyze the factors that influence a player's possession statistics.

Factors That Affect Possession

Like plus/minus, all possession stats are simply counts of what happened on the ice while a player was there, rather than any player's individual actions. In many ways, this is their strength: it means that if a player does important things that we forgot to count, they will show up in the results. There are no NHL stats for "good positional play" or "excellent clearing pass," but these things will show up, eventually, in shot metrics.

However, this also means that a player's stats are subject to the slings and arrows of outrageous fortune and affected by factors well beyond his control. There are some major ones that we can control for.

Adjusting for Team Quality

In 2014–15, Jarret Stoll and Nathan MacKinnon posted similar USAT results. Stoll's Kings had a USAT% of 50.6% while he was on the ice, while MacKinnon's Avalanche had a USAT% of 49.5%. And yet, despite

Stoll's slightly superior results, we know that MacKinnon's performance was far superior. Why?

Among those who follow advanced statistics, the standard measure of possession stats such as SAT is not raw SAT but SAT rel, also known as Corsi rel. SAT rel, which means "SAT relative," is a player's SAT adjusted for their team, so that the average SAT rel on any team is zero. This is an eminently reasonable adjustment: as we saw with plus/minus, all possession stats are highly team skewed. As an example, here are the league leaders in 2014–15 in raw USAT.

LEAGUE LEADERS IN UNBLOCKED ATTEMPTED SHOT DIFFERENTIAL (USAT), 2014–15*

NAME	TEAM	RAW USAT	NAME	TEAM	RAW USAT
Drew Doughty	LAK	+284	Mike Ribeiro	NSH	+211
Jake Muzzin	LAK	+278	John Tavares	NYI	+204
Nick Leddy	NYI	+264	Jeff Carter	LAK	+201
Joe Thornton	SJS	+262	Mattias Ekholm	NSH	+198
Anze Kopitar	LAK	+260	Anton Stralman	TBL	+198
Johnny Boychuk	NYI	+245	Eric Staal	CAR	+190
Joe Pavelski	SJS	+233	Sidney Crosby	PIT	+186
Marian Gaborik	LAK	+226	Kris Letang	PIT	+186
Filip Forsberg	NSH	+221	Vladimir Tarasenko	STL	+185
Patrice Bergeron	BOS	+214	Chris Kunitz	PIT	+184

[* Shooting percentage data obtained from the NHL's website, accessed August 31, 2015, http://www.nhl.c om.]

They are all good players in their own right, but we see there are five Kings in the top 20, three Islanders, three Penguins, three Predators, and two Sharks. Even more tellingly, among the top six, Muzzin was paired with Doughty for over two-thirds of his ice time, and they often played behind Kopitar as well, while Boychuk played with Leddy 83% of the time. This is hardly an individual stat.

How does this compare to the league leaders in team-adjusted USAT?

LEAGUE LEADERS IN TEAM-ADJUSTED USAT, 2014–15*

NAME	TEAM	ADJ USAT	NAME	TEAM	ADJ USAT
Joe Thornton	SJS	232.5	Anze Kopitar	LAK	125.3
Joe Pavelski	SJS	201.5	Henrik Sedin	VAN	125.0
Patrice Bergeron	BOS	192.9	Filip Forsberg	NSH	124.4
Brad Marchand	BOS	153.7	Jason Demers	SJS/DAL	123.5
Michael Raffl	PHI	152.4	Patrick Wiercioch	OTT	123.1
Gabriel Landeskog	COL	146.6	Brendan Gallagher	MTL	121.8
Nathan MacKinnon	COL	140.0	Nick Leddy	NYI	121.4
Scott Hartnell	CBJ	134.7	Eric Staal	CAR	120.2
Martin Erat	ARI	131.4	Jakub Voracek	PHI	117.4
P.K. Subban	MTL	129.3	Pavel Datsyuk	DET	116.0

[* Shooting percentage data obtained from the NHL's website, accessed August 31, 2015, http://www.nhl.com.]

This list is significantly different, although we can see that some players have still arrived in pairs. Thornton and Pavelski played together 83% of the time, and they're so dominant separately that, together, they were unstoppable. Bergeron is so good that anyone who plays with him 80% of the time, as Marchand did, will see his numbers improve as well. Similarly, the Sedins are practically inseparable, so they will either make it together or not at all. But this list seems a lot more representative than the previous one: there are players from 13 different teams, and more of the players like MacKinnon, Subban, Eric Staal, Voracek, and Datsyuk, who are primary drivers of possession on their teams.

Overall, 53% of a player's USAT can be explained by the team he plays on. This is huge and means that unadjusted numbers are almost meaningless. On the flip side, adjusting for team isn't perfect either. After all, better teams have better USAT numbers because they have better players, right?

This is true at a team level but not at an individual level. For example, let's say I want to know how much better Jonathan Toews is than Average McMedium. The ideal set-up would be to put both Toews and McMedium on a team of average players and see how well each of them does. We expect Toews to do better than McMedium, obviously, and the difference between their results will be directly due to their own performance, since everyone else around them is the same. However, if we assemble an all-star team of 12 Toewses and six Duncan Keiths and pit them against a team of 18 McMediums, each of the Toewses on the all-star team will do *much* better than the lone Toews on the average team because their linemates are so much better.

After adjusting for team, we can see that MacKinnon was, indeed, far superior to Stoll.

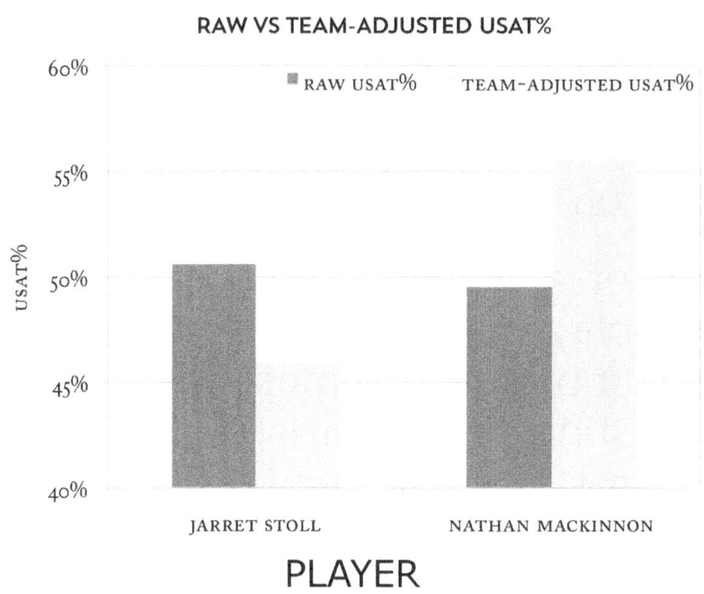

One final comment: adjusting for a player's specific linemates is obviously better than adjusting for the team as a whole, since it guarantees that we are adjusting for the benefit the player actually got. However, it requires calculations that are significantly more complex than simply adjusting for the team, which can be done by anyone with raw data and a spreadsheet. If you have a data source that adjusts by shared ice time instead of at the team level (or if you are inclined to perform these calculations yourself) by all means use it. It will be slightly more accurate than a team-level adjustment.

Adjusting for Teammates

Players on better teams get to play with better linemates and get better results because of this—that part is fairly obvious. The next level is figuring out if some players get all the better linemates on their team. After all, playing for Boston will help you a lot if you play with Patrice Bergeron but not as much if you play with Gregory Campbell. So the next, obvious step is to adjust for the quality of one's linemates, not just of the team.

This is not as obvious at it seems. When players play together a significant portion of the time, there is simply no way to disentangle their results. For example, in 2014–15 Corey Perry had Ryan Getzlaf as his centre for 79% of his ice time; this means he played only 200 even-strength minutes without Getzlaf during the entire season. Unless one of the two players is dramatically better than the other, there will be no valid statistical way to disentangle drivers from passengers in this arrangement. We must simply accept that we are judging the results of these players as a unit.

What we see, in practice, is that the distribution of linemates spreads the wealth pretty evenly, except for a few outliers. This is mainly because, while forward lines and defenceman pairings are often quite stable, forwards and defencemen get mixed on a regular basis, which means that both first-liners and fourth-liners get to play with both number-one and number-six defencemen and vice-versa.

Let us take as a case study the 2014–15 Los Angeles Kings, who had the highest raw shot differential of any team that season. The players with the strongest linemates were Anze Kopitar and Marian Gaborik, who had the benefit of playing with each other; when Gaborik was on the ice, Kopitar was there 86% of the time. On the flip side, the player with the weakest linemates was Kyle Clifford, who played most often with Jordan Nolan but had no fixed linemates for the entire season. Here we see the players' raw USAT%, team-adjusted USAT%, and teammate-adjusted USAT%, and the difference between the latter two (diff).

LOS ANGELES KINGS INDIVIDUAL POSSESSION METRICS, 2014–15*

PLAYER	RAW USAT%	TEAM-ADJUSTED USAT%	TEAMMATE-ADJUSTED USAT%	DIFF
Marian Gaborik	59.2%	55.3%	54.6%	-0.8%
Anze Kopitar	58.9%	55.1%	54.4%	-0.7%
Jake Muzzin	57.6%	53.7%	53.4%	-0.3%
Jeff Carter	56.6%	52.8%	52.7%	-0.1%
Trevor Lewis	56.6%	52.7%	52.6%	-0.1%
Tyler Toffoli	56.4%	52.6%	52.6%	0.1%
Brayden McNabb	56.1%	52.2%	52.3%	0.1%
Drew Doughty	55.9%	52.1%	51.9%	-0.1%
Justin Williams	55.7%	51.9%	51.9%	0.1%
Tanner Pearson	55.4%	51.6%	51.8%	0.2%
Dustin Brown	54.2%	50.4%	50.4%	0.0%
Matt Greene	54.1%	50.2%	50.2%	0.0%
Dwight King	53.3%	49.4%	49.5%	0.1%
Kyle Clifford	53.3%	49.4%	50.2%	0.8%
Alec Martinez	52.8%	48.9%	49.2%	0.3%
Jarret Stoll	50.6%	46.7%	46.7%	0.0%
Robyn Regehr	50.1%	46.2%	46.5%	0.2%
Mike Richards	49.2%	45.4%	45.7%	0.3%
Jordan Nolan	47.3%	43.4%	44.0%	0.5%

Note: Only players who played a minimum of 400 minutes are included.

[* Raw possession data obtained from the NHL's website, accessed August 31, 2015, http://nhl.com.]

As predicted, Kopitar and Gaborik see the biggest downward adjustments, while Clifford sees the biggest upward adjustment. Given the spread of adjusted USAT%, 0.8% is certainly large enough that it needs to be accounted for, although for most players it is far smaller than that.

Across the league, the players who saw the biggest downward adjustments were the ones who played for strong units that stayed together for most of the season. Melker Karlsson played only 53 games for the Sharks, but when he did play he was paired mostly with Pavelski and Thornton, while Reilly Smith was the most common linemate for Bergeron and Marchand. These two trios, plus the inseparable Sedins in Vancouver, get eight of the top ten spots.

10 BIGGEST NEGATIVE TEAMMATE ADJUSTMENTS IN USAT%, 2014–15*

PLAYER	TEAM	RAW USAT%	TEAM ADJUSTED USAT%	TEAMMATE ADJUSTED USAT%	DIFF
Melker Karlsson	SJS	54.4%	53.6%	52.2%	-1.3%
Reilly Smith	BOS	54.1%	53.5%	52.3%	-1.2%
Brad Marchand	BOS	55.9%	55.3%	54.2%	-1.1%
Joe Pavelski	SJS	57.2%	56.4%	55.3%	-1.1%
Joe Thornton	SJS	58.7%	57.8%	56.9%	-1.0%
Daniel Sedin	VAN	53.7%	53.4%	52.4%	-1.0%
Nathan MacKinnon	COL	49.5%	54.6%	53.7%	-0.9%
Henrik Sedin	VAN	54.7%	54.4%	53.6%	-0.8%
Marian Gaborik	LAK	59.2%	55.3%	54.6%	-0.8%
Patrice Bergeron	BOS	57.3%	56.7%	55.9%	-0.8%

[* Raw possession data obtained from the NHL's website, accessed August 31, 2015, http://nhl.com.]

Among the players with the biggest positive adjustments, we see two types: weak units that play together regularly and players on strong teams that don't get to play with the superstars. It's not fair to penalize Andrew Desjardins, Daniel Paille, and Gregory Campbell for playing on a strong team if they never get to play with Pavelski or Bergeron. Brad Malone,

Patrick Dwyer, and Jay McClement formed a defensive unit for Carolina that got weak results all season but rarely had the benefit of playing with strong possession drivers like Eric Staal.

10 BIGGEST POSITIVE TEAMMATE ADJUSTMENTS IN USAT%, 2014–15*

PLAYER	TEAM	RAW USAT%	TEAM-ADJUSTED USAT%	TEAMMATE-ADJUSTED USAT%	DIFF
Jared Boll	CBJ	36.0%	38.5%	40.6%	2.2%
Andrew Desjardins	SJS	41.3%	40.5%	42.3%	1.9%
Brad Malone	CAR	46.3%	44.5%	46.0%	1.5%
Daniel Paille	BOS	45.7%	45.1%	46.4%	1.3%
Jim Slater	WPG	45.2%	43.6%	44.8%	1.2%
Gregory Campbell	BOS	41.5%	40.9%	42.0%	1.2%
Patrick Dwyer	CAR	45.7%	43.9%	45.0%	1.1%
Jay McClement	CAR	44.9%	43.1%	44.2%	1.1%
Luke Glendening	DET	44.5%	42.9%	43.9%	1.0%
Josh Gorges	BUF	35.5%	44.8%	45.9%	1.0%

[* Raw possession data obtained from the NHL's website, accessed August 31, 2015, http://nhl.com.]

You will notice that all the players with positive adjustments still have a teammate-adjusted USAT% significantly under 50%, while all the players with negative adjustments still end up over 50%. This comes back to my original point: it is very difficult to separate the contributions of linemates who play regularly together. By definition, having a big negative teammate adjustment requires two things: that you played regularly with the same players, and that you achieved strong results while playing together. By this logic, it seems obvious that "good results" and "good teammates" will go hand in hand, since your linemates will share your good results.

While the choice of teammates has a measurable impact for some players, for most it is negligible. Overall, which players on your team you play with accounts for only 3% of USAT on average.

Correct teammate-adjusted USAT% is close to the best we can do in terms of a shot differential indicator that is not affected by who a player plays with, but it is important to remember that the data has limits. If a player is injured, it could throw the line pairings into disarray, weakening everyone's results while that player is injured; the observed result will be to assume that he was the glue holding everything together, when it's really the effect of the sudden absence that we are seeing.

Adjusting for teammates is the first essential step we must do before we can interpret any possession metrics in a useful way, but there is also a second critical, and more contested, adjustment to do.

Adjusting for Zone Starts

Why does Manny Malhotra still have a job? After all, here is a player who has never scored more than 35 points in a season, despite getting a decent amount of ice time. Our advanced stats are supposed to help us identify two-way players who do more than score points, but in 2014–15 Malhotra got only a paltry 36.5% USAT% in 58 games with the Montreal Canadiens, making him one of the worst possession forwards in the entire NHL. Poor Malhotra seems bad no matter which metric we use.

Of course, true advanced stats fans know the answer: Malhotra is almost exclusively used in his own zone, giving him limited opportunity for offensive production or even to generate shots on the opposing net. In fact, Malhotra is the poster boy for a statistic known as zone starts.

Zone starts, or offensive zone start percentage, was another innovation of Ferrari and Desjardins. The basic concept of zone starts is extremely simple: a team that is taking a faceoff in the offensive zone is advantaged in the shot differential game. If they win the faceoff, they are likely to get a shot on goal shortly; if they lose the faceoff, the defensive team still needs to break out of their zone and bring the puck to the other end unmolested before they can attempt any kind of offence. The raw data confirms this insight, and it was useful in a previous chapter on searching for the league's best at faceoffs.

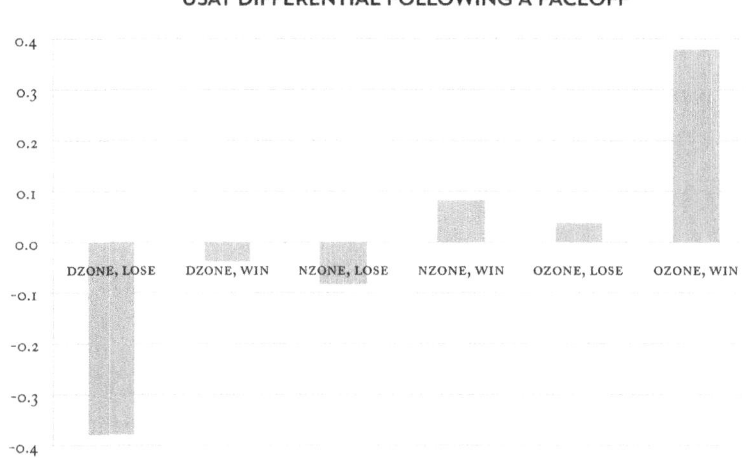

There is a second, important element to this: even on the same team, some players get more offensive faceoffs than others. These players will see all of their numbers—USAT, plus/minus, goals, assists, shots on goal—improved by this deployment. It is important to correct for this if we are to analyze each player's contributions to his team fairly.

A lot of analysis has been done on the impact of zone starts on USAT. Early hockey statisticians used to simply run a correlation between raw Corsi and raw zone starts, and they came to the conclusion that zone starts had a *huge* impact on Corsi. This is wrong, however, because some players play for bad teams and others play for good teams. The bad teams have many defensive zone starts and many shots against because they are bad; in fact, many shots against that don't lead to a goal lead to a faceoff in your own zone. If you look at everyone on the 2014–15 Sabres, they all have negative zone stats and negative Corsi, but all that is due mostly to the team effect, discussed already.

A few years ago, Eric Tulsky produced what I believe is still the best overview of the impact of zone starts on USAT, and he concluded that each extra offensive faceoff a player took led to an average increase in his USAT of 0.3.* This matches my own calculations, and I believe this should be the roughly accepted value for any kind of zone start correction.

[* Eric Tulsky, "Review: Zone Start Adjustments to Shot Differential," *NHL Numbers* (blog), November 5, 2012, http://nhlnumbers.com/2012/11/5/zone-start-adjusted-corsi-corrections-faceoffs.]

Just the raw numbers of what happens after a faceoff (see previous chart) tell us a lot. We can see that winning or losing the faceoff has a huge impact, but since virtually nobody in the NHL wins more than 60% or less than 40% of his faceoffs, all players will see a mix of wins and losses.

In the last couple of years, there has been a pushback against zone start adjustments by analysts such as David Johnson, Matt Cane, and Nick Abe, on the assumption that they don't matter as much as we thought.* I believe this is a mistake. Anytime we are not adjusting correctly for data that has an impact, we are doing ourselves a disservice, and we are lessening our ability to maximize the value we get out of our data. I guarantee that an NHL general manager trying to understand how well his players are performing would want to adjust for zone starts, not simply ignore them on the assumption that the effect is small.

[* David Johnson, "Why Zone Starts Don't Matter Much," *Hockey Analysis* (blog), December 13, 2014, http://hockeyanalysis.com/2014/12/13/zone-starts-dont-matter-much/; Matt Cane, "How Much Do ZoneStarts Matter Part I: (Maybe) Not as Much as We Thought," *Puck Plus Plus* (blog), January 15, 2015, http://puckplusplus.com/2015/01/15/how-much-do-zone-starts-matter-i-maybe-not-as-much-as-we-thought/;Nick Abe, "Why I Don't Adjust the Data for Zone Starts," *Xtra Hockey Stats* (blog), March 29, 2015,http://xtrahockeystats.com/wordpress/?p=17. Nick Abe, "Why I Don't Adjust the Data for Zone Starts," *Xtra Hockey Stats* (blog), March 29, 2015, http://xtrahockeystats.com/wordpress/?p=17.]

To illustrate how important zone starts can be, here is a chart of the 10 players with the largest positive and negative adjustment on their USAT for 2014–15, using an adjustment value of 0.29 USAT per faceoff.

Zone starts don't justify everything. Filip Forsberg and Jakub Voracek, who got very favourable zone starts, still do well after the adjustment, while our hero Manny Malhotra is still negative after the adjustment but no longer appears so bad.

Once we have eliminated team effects, zone starts explain 12% of a player's teammate-adjusted USAT. This means that they explain about 5.4% of the overall USAT, leaving only 39% of raw USAT explained by a player's individual performance, luck, and other factors.

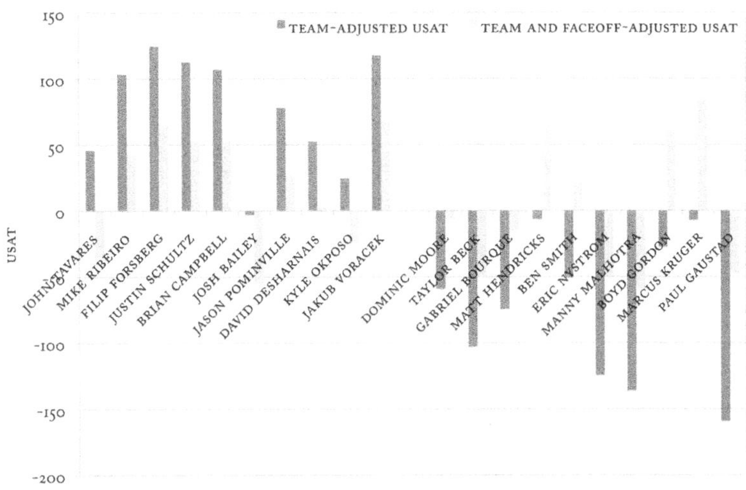

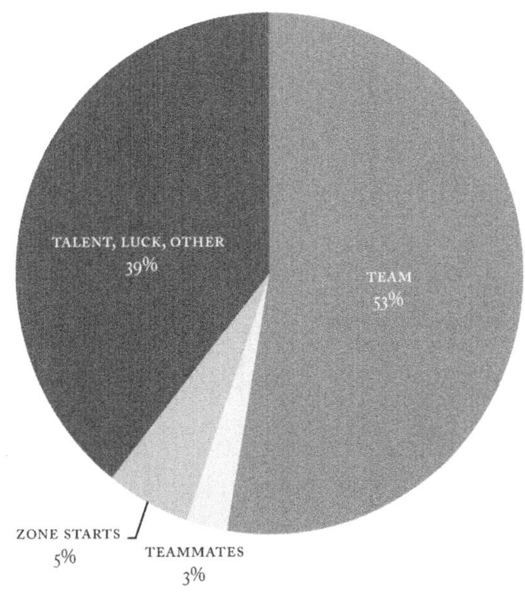

CASE STUDY: *Who Is Better, Thomas Vanek or Matt Moulson?*

In October 2013, Vanek was traded from the Buffalo Sabres to the New York Islanders for Moulson, a first-round pick, and a second-round pick. While they are only two months apart in age, these two players had very different paths into the NHL. Vanek was a can't-miss talent who was drafted fifth overall by the Sabres in the 2003 entry draft. He broke into the NHL two years later, and by his second season he was a 43-goal scorer, leading the NHL in plus/minus and helping the Sabres to the conference final. Moulson was drafted the same year, 2003, but 263rd overall and by the Pittsburgh Penguins; to put this in perspective, there are only 210 picks in the NHL draft today. It would have been unsurprising had he never played a game in the NHL, but he persevered. He was signed by the Los Angeles Kings in 2006, playing three years for their AHL farm team and 29 games in the NHL before being signed by a desperate Islanders team that had finished last overall in 2008–09. Moulson scored 30 goals in his first full season with the Islanders and never looked back.

At the time of the trade, Vanek was still considered the franchise player for the Buffalo Sabres, while Moulson was an established scorer who had strung together three 30-goal seasons in a row before having a down season in 2012–13. Both were free agents at the end of the season, so both moved a lot: Moulson was traded to the Minnesota Wild before re-signing with the Sabres at the end of the 2013–14 season, while Vanek was a rental for the Montreal Canadiens before signing as a free agent with the Wild. At the time of the initial trade, Vanek was considered a far more valuable player than Moulson, but with the benefit of hindsight, we can ask who has been better?

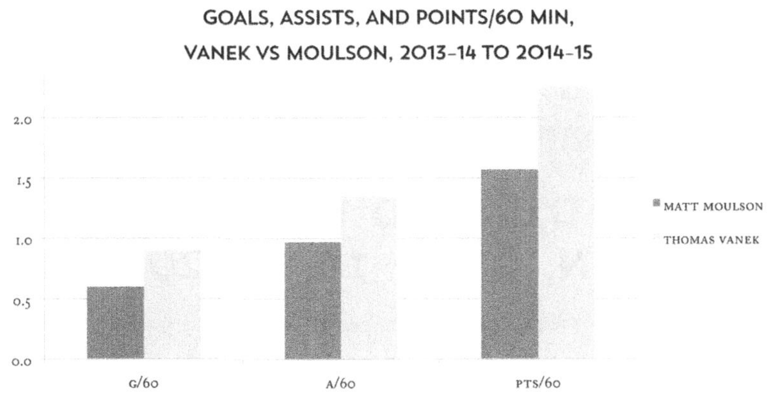

Let's start with the headline numbers, looking just at five-on-five play for now. Over the two seasons, Vanek scored 32 goals and 48 assists in 2,130 minutes, while Moulson scored 21 goals and 34 assists in 2,104 minutes. So Vanek was scoring at a relatively elite 2.25 points per 60 minutes, while Moulson achieved just 1.57 points per 60 minutes, slightly below the level for a top-six forward.* Advantage Vanek, by a large margin. We can see why Vanek was considered far more valuable than Moulson.

[* Rob Vollman, "Howe and Why: Top-Six Forwards, Part 1," *Hockey Prospectus* (blog), December 10, 2009, http://www.hockeyprospectus.com/puck/article.php?articleid=387.]

What about the on-ice numbers? Both Moulson and Vanek had pretty weak raw possession numbers, in particular Moulson, who spent the majority of both seasons with the Sabres, the worst team in the league. Vanek's raw USAT% was 46.5% over the two seasons, while Moulson's was an even worse 44.3%. However, we just saw in the previous sections that team and zone starts have a massive effect on a player's USAT numbers, and Moulson in particular is a candidate for a significant adjustment given the weak teams he has played for.

RAW AND ADJUSTED USAT%, VANEK VS MOULSON, 2013–14 TO 2014–15*

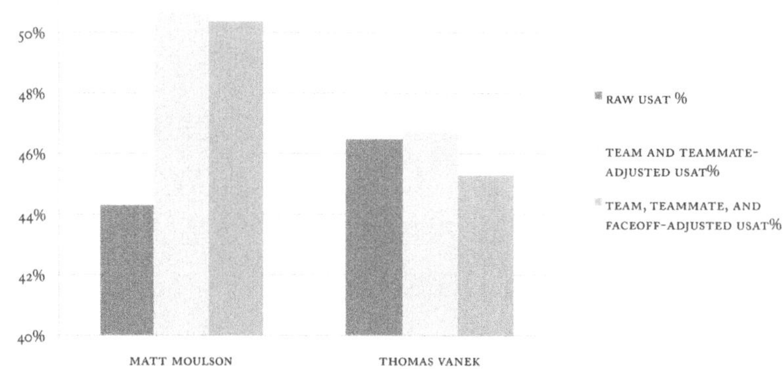

[* Raw USAT% data obtained from the NHL's website, accessed August 31, 2015, http://nhl.com.]

We see that, after adjusting for team and teammates, Moulson's USAT% number bumps just above 50%, to 50.7%, while Vanek's barely improves at all, to 46.7%. Remember that Vanek spent all of 2014–15 with the Wild, one of the league's top 10 possession teams.

The zones in which their faceoffs occurred add another layer of adjustment to these numbers. Both Moulson and Vanek are offensive players: Moulson played 193 minutes on the power play in 2014–15 and only 11 minutes short-handed, while Vanek played 232 minutes with the man advantage but only 24 seconds short-handed. Vanek is particularly one-dimensional and is used as such by his coaches, being deployed in the offensive zone whenever possible. Adjusting for zone starts reduces Moulson's USAT% to 50.4% and Vanek's down to 45.3%, one of the lowest numbers for a first-line player in the entire league.

So what am I trying to tell you? That Vanek is bad? Absolutely not. Our original observations about points per 60 minutes still hold, notwithstanding the possession numbers. Vanek's career shooting percentage in the NHL is 14.5%, a stratospheric number that shows he is capable of capitalizing on offensive opportunities at a rate much higher than regular NHLers. His teams' on-ice shooting percentage at even strength over these two seasons was 10.2% versus the NHL average of 7.7%, and we know in his case that this is no fluke; these were not even two of his best seasons in the NHL. He is a liability on possession but an asset on percentages, with the net benefit being slightly positive at even strength. Vanek's team-adjusted plus/minus is +75 in 743 games since he entered the NHL. On top of that, Vanek's contibution to the power play during his better years was massive: he led the entire NHL in power-play goals in 2008-09. The best parallel for Vanek is Ilya Kovalchuk, another pure scorer who looked terrible based only on possession numbers.

Moulson, meanwhile, is more of a "traditional" NHL first or second-liner (although he was/is a first-liner on the Islanders and Sabres, teams without much first-line offensive talent), a player who is at least neutral on possession while providing some offensive advantages. His career team-adjusted plus/minus is +75, the same as Vanek, over 474 games, indicating that he, too, is an asset overall at even strength. In Moulson's case, his teams' on-ice shooting percentage was 8.6%, slightly better than the NHL average.

The lesson is that there is no one stat to analyze players. Adjusted USAT% is very informative and tells us a lot about a player's value, but percentages can be misleading because they are very noisy in the short term. For some players like Thomas Vanek, however, their entire value shows up mainly in the percentages, so we cannot discount those, but we must ensure we are not drawing conclusions based on too small a sample size.

Adjusting for Quality of Competition

The last big factor affecting SAT is quality of competition (QoC). Since the beginnings of Corsi, QoC together with quality of teammates (QoT) and zone starts have been considered the three big factors affecting a player's possession metrics.

That quality of competition would influence a player's results is obvious. If you play against better players, you will do worse; otherwise they're not really better players. So the debate has centred not on whether QoC has an impact, but on how much of an impact it has.

One challenge is that, unlike other usage metrics, quality of competition is highly correlated with player talent. Better players get matched up against better players. This is a typical usage model in the NHL that has been baptized power versus power. But if strong players play against strong players and have better possession numbers *and* higher QoC, how can we tell if QoC even matters?

It's very difficult. Early studies, like this one by Desjardins, seemed to indicate that yes, it matters.* In recent years, more people have gravitated toward the view that the spread of QoC is small enough that it can be ignored. Recent studies, including one by Domenic Galamini, seem to confirm this view.**

[* Gabriel Desjardins, "Further To: Does QualComp Matter?," *Arctic Ice Hockey* (blog), July 27, 2011, http://www.arcticicehockey.com/2011/7/27/2294013/further-to-does-qualcomp-matter.]

[** Domenic Galamini, "The Significance of Quality of Competition," *Blue and White Brotherhood* (blog), July 30, 2014, http://blueandwhitebrotherhood.blogspot.ca/2014/07/the-significance-of-quality-of.html.]

There are many ways of measuring QoC. The earliest measures, available on Gabriel Desjardins' *Behind the Net,* were plus/minus based. As shot-based metrics became more widespread, this evolved to Corsi-based QoC and then relative Corsi QoC. Currently, the most widespread measures of QoC are calculated using either SAT relative or ice time.

Each measure has its benefits. In terms of measuring which players have the most difficult assignments, ice time – or plus/minus–based measures are usually the best. As we saw in previous sections, there are many offensive players like Thomas Vanek or Alexander Ovechkin who don't put up strong possession numbers but are certainly considered top competition. These players also tend to get a lot of ice time, so ice-time QoC will see them as difficult assignments. From the point of view of adjusting SAT for quality of competition, however, SAT-relative measures are definitely the best. In practice, all measures tend to give similar results. The correlation between ice-time QoC and SAT-relative QoC was 0.80 for all NHL players with 800+ even-strength minutes in 2014–15,* and Rob's famed player usage charts use SAT relative as their metric of choice.

[* Raw data for that calculation obtained from *Hockey Abstract,* accessed August 31, 2015, http://www.hockeyabstract.com/testimonials/nhl2014-15playerdata.]

This leaves the ultimate question: assuming we are happy to use quality of competition measures based on SAT relative, how much impact does this ultimately have on a player's results? The best analysis I have seen on this subject comes, once again, from Eric Tulsky, and if you want to understand the impact of QoC you could probably read his article alone and be up to speed. Tulsky showed two things. First, who you play against on a given shift matters a lot: SAT results differ by huge margins between strong and weak opponents. Second, overall differences in QoC between players are extremely small, small enough that they are drowned out by the noise.* Unlike zone starts or linemates, where a coach has complete control over who he puts on the ice in any given situation, quality of competition depends on the other team as well. Half the time, you don't get the last line change anyhow, so you have no control over the matchup. I believe the other reason is that defensive zone choices take precedence over matchup choices. If I'm Peter Laviolette, the Predators coach, and there's a faceoff in our zone and Paul Gaustad is available, there are good chances that I'll put him on the ice, regardless of whether the other team has put out its first line or its third line.

[* Eric Tulsky, "The Importance of Quality of Competition," *NHL Numbers* (blog), July 23, 2012, http://nhlnumbers.com/2012/7/23/the-importance-of-quality-of-competition.]

The bottom line is that you should adjust SAT by quality of competition if you have the data to do so, since there's really no reason not to. But unlike zone starts or teammates, it really won't affect the results by much, and if your data source doesn't compensate for QoC there's no reason to worry about the conclusions you'll draw. The real value in quality of competition measures is in seeing which players are being given the responsibility to shut down the opposing team's best players. Quality of competition is a window into a coach's level of confidence in his players, and while this is not always 100% reliable it is another data point that isn't correlated with on-ice results. Players with high QoC measures can be two-way superstars like Jonathan Toews, old masters like Jarome Iginla, or purely defensive forwards like Brooks Laich. There is also no real correlation between offensive role and quality of competition among defencemen. Mark Giordano and Josh Gorges don't have much in common in terms of offensive ability, but both of them are among their team's leaders in QoC.

How Reliable Is SAT?

I have covered all the elements that affect players' possession numbers and how to adjust for them to get a more meaningful measure. At the end of all this, we must ask how useful is the number we have produced? An ideal statistic would be completely independent of context and would represent some actual level of talent or performance. Shootout statistics, for example, are not affected by a player's teammates, but they are so noisy that they don't represent much unless you have several years of data to work with. At the other end, unadjusted SAT certainly represents true ability, since a team's results are fairly consistent over a single season, but it is drowned by context for individual players.

The bottom line is that a good hockey statistic must meet two criteria: it must not be overly affected by a player's usage, be it the team he plays for or the amount of ice time he gets, and it must be somewhat correlated from year to year, indicating that it represents true skill and not random chance. A good example of this is even-strength points per 60 minutes (pts/60). Unlike raw points, it is not affected by the amount of ice time a player gets, filters out the power play, and has enough information that it is somewhat meaningful over the course of an entire season. It is of course affected by the quality of a player's linemates, so it is far from perfect.

The year-over-year correlation of pts/60 for forwards between 2013–14 and 2014–15 was 0.59, indicating pretty high correlation but far from perfect (for defencemen it was much lower, at 0.39, since defencemen score so few points at even strength).

EVEN-STRENGTH POINTS/60 MINUTES, FORWARDS, 2013–14 VS 2014–15*

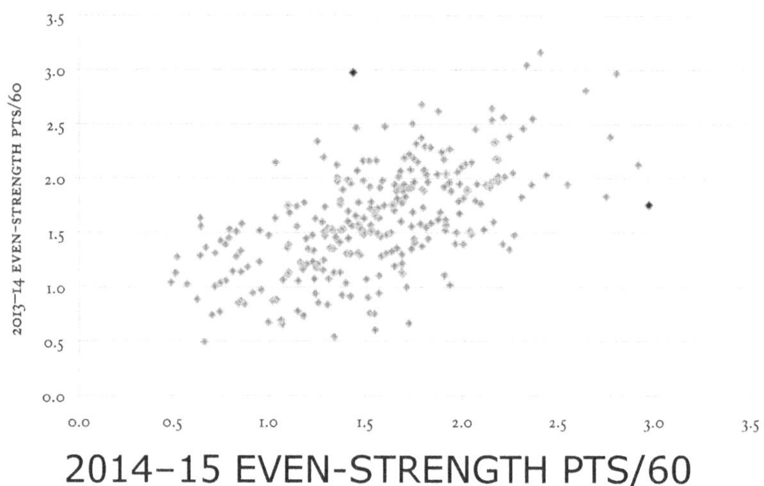

[* Raw points data obtained from the NHL's website, accessed August 31, 2015, http://nhl.com.]

Obviously players do change, so we would not expect even a perfect statistic to be constant from season to season. The player on the far right of the above chart (in black), who scored only 1.8 pts/60 minutes in 2013–14 but 3.0 in 2014–15, is Tyler Johnson, who clearly improved as a player between the two seasons; better linemates only helped his situation. On the flip side, Gustav Nyquist (top centre, also in black) scored 3.0 pts/60 minutes in 2013–14 and had Wings fans believing he was the NHL's next superstar. In 2014–15, as predicted by those who pay attention to percentages, he crashed back to Earth with only 1.4 pts/60 minutes.

The year-to-year correlation for adjusted USAT/60 is similar, at 0.54. This means that we will expect to see some cases like Johnson or Nyquist, where players go from being one of the best in the league one season to middle of the pack the next. The player who suffered the biggest drop-off from 2013–14 to 2014–15 was Ryan Stanton, who went from a +7.1 USAT/60 to −4.7 (top left, in black, on the following graph), while the player with the greatest improvement was Patrick Wiercioch, who went from −0.9 to +7.0 (far right, in black).

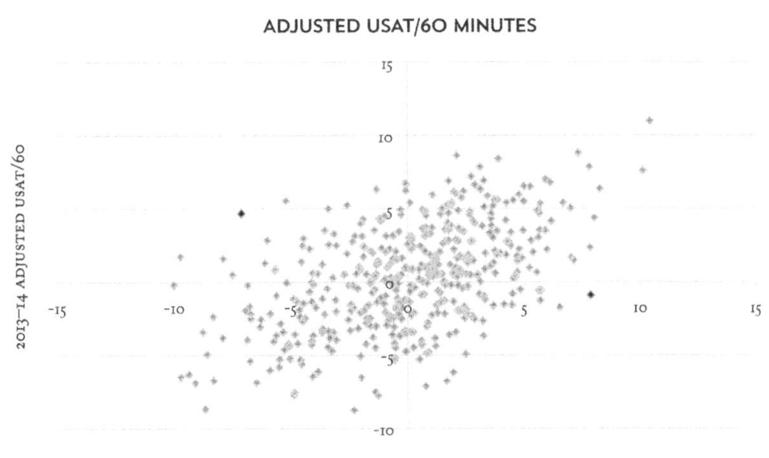

2014–15 ADJUSTED USAT/60

However, these are clearly outliers. The vast majority of players had similar performances in 2013–14 and 2014–15. We can see Patrice Bergeron, who led the league in both seasons (11.0 in 2013–14 and 10.4 in 2014–15), at the top right of the graph, mocking everyone else, while players like Anze Kopitar (8.8 in 2013–14 and 7.3 in 2014–15) and Joe Thornton (7.6 and 10.1) also put up solid results year after year. Adjusted USAT seems to meet our two criteria: it must not be overly dependent on context, and it must have some skill-based component such that the numbers have some year-to-year correlation. Like pts/60, we can see that it also contains some noise, so we should be wary about using numbers over a short period of time, especially when they are very surprising or go against a player's established track record.

Closing Thoughts

In this chapter, I've tried to summarize the important insights that have emerged during the development of shot-based metrics. Shot-based metrics, also known as possession metrics, emerged as a substitute for plus/minus. Possession numbers are more useful than plus/minus because they have a large sample size, which allows us to draw conclusions from a smaller number of games than would otherwise be the case. They are also useful because, from the point of view of an individual player, they eliminate the impact of goaltending, which would otherwise bias the results, especially over the short term. Finally, because of their large sample size, they are persistent: possession numbers tend to hold over the course of a season, and, consequently, it is more accurate to estimate a team's future success from its shot differential than from its goal differential, especially if the estimate is based on less than a full season's worth of data.

Possession statistics are interesting because they can be calculated both at the individual and team levels. At the individual level, they account for much of what a player contributes at even strength but not for a player's finishing talent, which can be a major contributor for elite forwards. However, individual players' results are highly influenced by the team they play for and whether they are used primarily in an offensive or defensive role. It is important to correct for these usage factors in order to get a valid idea of how well a player is performing.

Most importantly, possession is the biggest *repeatable* factor in the success of players and teams in the NHL today. As such, they have become one of the essential tools in measuring players and teams. Now that they have been officially adopted by the NHL, their mainstream use will continue to grow in the near future.

WHAT WAS THE MOST ONE-SIDED TRADE OF ALL TIME?

by ROB VOLLMAN

Consider this: Boston was set up for a decade after acquiring Phil Esposito from Chicago, Colorado's mini-dynasty began shortly after trading Eric Lindros to Philadelphia, and Wayne Gretzky's trade to Los Angeles changed the landscape of NHL hockey forever. Comparing trades like these to determine which ones were the most lopsided of all time provides great fodder for fans' arguments in the sports bars across the nations.

Given how emotional these debates can be, and how often they're grounded in personal impressions, some cold, hard statistics can be just the thing to help sort out one deal from another. How good was each player, really? What kind of impact did he make on his new team and throughout the remainder of his career?

The basic requirement for this type of objective analysis is to have a catch-all statistic that measures all of a player's offensive and defensive contributions with a single number, like Tom's goals versus threshold (GVT). By adding up the GVT of every player involved in a trade, starting from the moment it was executed until the involved players retired, it is possible to quantify who won each trade and by how wide a margin.

Even when using an objective measurement such as GVT, several different ways remain to examine each trade. For example, should a player's entire future career be considered or only the portion spent with his new club? If a draft pick is involved, should it be given an expected value based on its overall position, or assigned the full value of whichever player was ultimately selected? How should individual and team awards as well as financial and contractual matters be factored in?

Let's briefly summarize the proposed approach, and the logic behind it, before diving into the top 20 most lopsided trades of all time.

The Objective Approach

There is no perfect approach to this sort of analysis; reasonable minds might view things differently. Every proposed analytic approach will have its own set of shortcomings while leaving plenty of ground for additional study, so it's best to be up front with the pros and cons of each.

GOALS VERSUS THRESHOLD

First of all, GVT isn't a perfect stat, but its strengths far outstrip its weaknesses, which should have only a negligible impact on this list. For example, GVT is explicitly designed for high-level comparison of players of different types and across different eras, but, in all fairness, it doesn't always measure defensive contributions perfectly. Neither does it measure non-quantifiable attributes, like leadership and clutch play. However, the total impact of GVT's failures on this list will ultimately prove to be minor.

INJURIES AND DEPARTURES

Of more pressing concern than GVT's shortcomings is the effect injuries and departures to other leagues can have on the evaluation of a trade. Several of these deals were lopsided only because a key player got hurt or left for Europe or the WHA, which are events that are largely out of a team's hands and unknown at the time of the transaction.

Luck can be a significant factor in some trades. For example, what would have happened if a 22-year-old Barry Pederson had fully recovered from shoulder surgery when he was dealt to Vancouver for a 20-year-old prospect named Cam Neely?

On the other hand, the calculations should be identical whether or not the acquired player sticks with his new team until retirement or is subsequently exchanged for another player down the line. For purposes of this analysis, what's important is which team left the bargaining table with the greatest package of future talent, whether it was ultimately squandered or multiplied or kept as is.

PROSPECTS AND DRAFT PICKS

The other significant issue in evaluating trades is the unpredictability of prospects and draft picks. Some deals wind up lopsided only because a particular prospect or draft pick turned out to be a Hall of Famer while similar picks and prospects turned out to be mediocre players, or perhaps didn't break into the NHL at all. The 1979 draft immediately comes to mind, where Edmonton snatched future Hall of Famers Mark Messier and Glenn Anderson with the third- and fourth-round selections obtained from St. Louis. The 1984 entry draft also comes to mind, when Montreal used Winnipeg's third-round selection to draft the most accomplished goaltender of all time, Patrick Roy.

That's why deals that were tilted by incredibly fortunate draft selections were set aside in this study. After all, you can't reward a team for making a shrewd trade when its success was dependent upon the rest of the league passing repeatedly on a particular player on draft day. However, these deals are included in the comprehensive reference table appearing near the conclusion of this chapter.

As the lone exception, teams trading away first-round selections shouldn't be given that free pass, especially if the picks were bound to be in the top 10. That kind of gamble warrants a big return and should definitely be considered lopsided if it does not. Likewise, teams should be responsible for determining a prospect's true value before shipping him off for some spare parts.

INDIVIDUAL AND TEAM AWARDS

Another deliberate omission in this proposed objective approach is that championships and Stanley Cups don't figure into the equations, nor do any individual player's performances in the postseason. For example, the New York Islanders won four Stanley Cups after acquiring Butch Goring from the Los Angeles Kings at the 1980 trade deadline, but how can we know if they wouldn't have done the same if they had retained Billy Harris and Dave Lewis instead?

Despite how important those details are emotionally, such factors are set aside in this analysis. While it subjectively makes perfect sense for teams in a critical season to acquire a player worth 10 GVT today in exchange for future players worth 20 GVT spread out over several seasons down the road, we don't have the insider information to make that judgment.

While it would certainly be interesting to assign a value for a Stanley Cup (or other important milestone) and to potentially decrease that value over time, that strikes me as overly arbitrary. Consider the approach used here as more of an analysis of who left the table with the greatest overall set of assets that day, and not whether the teams involved used those assets in the most effective and timely fashion.

FINANCIAL AND CONTRACTUAL MATTERS

Finally, contract-related matters are also ignored in this proposed objective approach. Even though many deals were triggered by a team's financial constraints, is that an excuse to make a lopsided trade?

While ignoring player salaries is a defensible choice for older historical trades, contract-related matters can no longer be ignored in the salary-cap era (2005 to present). Trading a superior player with a horrible contract for an inferior but value-priced player would actually be a great move, as would trading a pending unrestricted free agent for a value-priced player of far less talent.

Thankfully, any trades completed in the salary-cap era are too recent for us to determine how they will ultimately pan out for each team, and very few will be among those considered in this list.

So let's jump in.

The List

Using the objective approach just described, here are the top 20 most one-sided trades in history, including some of the logic behind them. A complete reference table with the statistics used in determining the order can be found at the end of this chapter.

1. **1978–79 NHL PRESEASON:**

    Los Angeles Kings trade first-round 1979 draft pick (Ray Bourque) to the Boston Bruins for Ron Grahame.

When the Bruins' GM Harry Sinden scooped up a first-round draft choice in exchange for some goaltending help, he had no idea that it would be used to acquire the most accomplished defenceman in history (according to career GVT), Ray Bourque. Consequently, it may appear unfair to rank this as the best trade of all time, but Sinden saw the strength of the upcoming draft class, found a middling team that was infamous for squandering its first-round selections in exchange for one more veteran (see trades 8 and 10), and selected the right goalie to move.

As for the Kings, when you trade away a first-round pick, you're taking a big risk. Sometimes you dodge a bullet, like when the Philadelphia Flyers traded two of them to acquire 24-year-old backup goalie Ken

Wregget, losing out on only rugged third-liner Rob Pearson and Steve Bancroft (six NHL games), but other times you give up franchise players like Guy Lafleur, Scott Niedermayer, or Ray Bourque.

While Bourque was undeniably the gem of the 1979 draft class, it was an outstanding year. Of the 12 forwards drafted in the first round, four topped a thousand points, including Mike Gartner, and of the nine defencemen, six played over 1,000 games. Even if Bourque had not been selected, this deal likely would have found a position high atop this list.

In fairness to Los Angeles GM George Maguire, the Kings were certainly in a pinch. They had lost their amazing goalie (and next GM) Rogie Vachon to Detroit in free agency, his aging backup Gary Simmons was most likely past the point of retirement, and they weren't sure if their young AHL goalie, Mario Lessard, was ready to handle the load (fortunately, he was). They desperately needed a goalie.

At the time, Ron Grahame seemed like just the ticket. The 28-year-old had just come off a great season for the Stanley Cup finalist Boston Bruins and had won the Hatskin Trophy as the WHA's best goalie in two of the preceding three seasons. Unfortunately, Grahame struggled in Los Angeles, to put it mildly, while the pick they gave up turned into a 19-time All-Star (wow!) and five-time Norris Trophy winner.

If there's a silver lining to this story, it's that Los Angeles did receive a package from Detroit in compensation for Vachon that ultimately included the draft choice that was used to select puck-moving Hall of Famer Larry Murphy (see trade 7). Of course, Maguire subsequently dealt him away for even less four years later (see trade 18), but let's not get too tangled up in that mess.

PLAYER	SEASONS	GP	G	A	PTS/GAA	GVT
Ray Bourque	22	1,612	410	1,169	1,579	492.4
Ron Grahame	3	74			4.37	-24.0

Bottom line, the Bruins acquired a valuable first-round selection for a goalie that appeared to be potentially among the league's best. It was a deal that didn't look bad on paper, but with a little bit of luck on draft day, it turned out to be statistically the most lopsided exchange in league history.

2. 1967 OFF-SEASON:

Chicago Black Hawks trade Phil Esposito, Ken Hodge, and Fred Stanfield to the Boston Bruins for Pit Martin, Jack Norris, and Gilles Marotte.

To quote Christopher Gasper of the *Boston Globe,* "there are no Big, Bad Bruins if the Blackhawks don't make a big, bad trade."* It was right before the expansion draft, and although an 18-year-old rookie named Bobby Orr had come to town, Boston had missed the playoffs for the eighth straight season. That's when new GM Milt Schmidt, who was technically still Hap Emms's assistant, made a deal that would forever change the face of his team.

[* Christopher Gasper, "Blackhawks Helped Start Bruins' Golden Era," *Boston Globe,* June 20, 2013, http://www.bostonglobe.com/sports/2013/06/19/milt-schmidt-recalls-landmark-trade-with-blackhawks/1CYrp9rQzqH3Wz0zyftHFJ/story.html

Even though it's always risky to trade prospects, swapping three for three ought to reduce the odds of a lopsided deal. Unfortunately for Chicago GM Tommy Ivan, this time it didn't. In his new role as a net-crashing goal scorer, Phil Esposito became one of the game's greatest superstars while Ken Hodge became a multi-time All-Star, and even Fred Stanfield was a solid top-six forward for years to come. This triggered an outstanding streak of 29 straight postseason appearances, including two Stanley Cups in the next five seasons.

At least Chicago didn't walk away empty-handed—Pit Martin was one of the league's top two-way centres throughout the early 1970s, and Gilles Marotte, while dealt away a couple of seasons later, was a useful puck-moving defenceman. Though unfortunate, the deal hardly crippled the franchise. Chicago struggled through two difficult seasons but then finished first eight times throughout the 1970s, which included two Stanley Cup appearances, and had a 28-straight postseason appearance run of its own.

PLAYER	SEASONS	GP	G	A	PTS	GVT
Phil Esposito	14	1,047	643	773	1,416	316.6
Ken Hodge	11	748	312	430	742	155.2
Fred Stanfield	11	807	201	393	594	89.5
BOSTON TOTAL	36	2,602	1,156	1,596	2,752	551.3
Pit Martin	12	990	288	452	730	131.5
Gilles Marotte	10	690	46	240	306	74.6
Jack Norris	3	35			4.04	−15.3
CHICAGO TOTAL	25	1,715	334	692	1,036	190.8

Interestingly, this wasn't the last great deal Boston made involving Esposito. Eight years later, GM Harry Sinden threw in Carol Vadnais and absolutely fleeced the New York Rangers for Brad Park and Jean Ratelle.

3. 1992 NHL ENTRY DRAFT:

Philadelphia Flyers trade Peter Forsberg, Steve Duchesne, Mike Ricci, Ron Hextall, Chris Simon, Kerry Huffman, 1993 first-round pick (Jocelyn Thibault), and 1994 first-round pick (Nolan

Baumgartner) to the Quebec Nordiques for Eric Lindros.

The drama behind one of the league's most infamous trades began when "the Next One," Eric Lindros, refused to sign with the Quebec Nordiques after being drafted, forcing the team into a situation where everybody knew he had to be moved. Normally, a situation as desperate as this is a recipe to get taken advantage of, and not the exact opposite.

PLAYER	SEASONS	GP	G	A	PTS	GVT
Peter Forsberg	13	708	249	636	885	206.0
Steve Duchesne	10	582	108	266	374	111.7
Mike Ricci	14	953	202	306	508	85.2
Jocelyn Thibault	14	586			2.75	60.5
Ron Hextall	7	327			2.71	51.0
Chris Simon	15	782	144	161	305	37.9
Nolan Baumgartner	10	143	7	40	47	8.4
Kerry Huffman	4	198	15	48	63	4.6
QUEBEC TOTAL	**87**	**4,279**	**725**	**1,457**	**2,182**	**565.3**
Eric Lindros	13	760	372	493	865	197.6

Quebec GM/coach Pierre Page and owner Marcel Aubut offered the vaunted prospect to the highest bidder, and did they ever get some incredible bids. Convinced that Lindros was a franchise player similar to Wayne Gretzky and Mario Lemieux, the Flyers left absolutely nothing off the table in their quest to acquire him, putting together the largest package ever offered for any collection of players, let alone a single one, ever.

In fairness, Lindros truly was an exceptional player, even though his career was cut short by repeated concussions throughout the 1999–00 season. He was a six-time All-Star and was arguably the league's best player for several seasons in his early 20s (when he was healthy, at least). Even his greatest detractor, Bobby Clarke, feels that his incredible success should have made him a "no-brainer" for the Hall of Fame.

Unfortunately, Flyers GM Russ Farwell simply gave up way too much in acquiring Lindros. In addition to superstar prospect Peter Forsberg, whose accomplishments essentially rival Lindros's, he gave up a prospect and two first-round draft picks, each of which resulted in useful NHLers, and four established NHLers, most of whom had many more years of productivity left. Oh, and $15 million with which to pay them all.

The Nordiques went from one of the league's worst teams to nine straight division championships and two Stanley Cups. While the Flyers did enjoy considerable success, they reached the finals only once (getting swept by Detroit), and they ultimately found themselves in the same boat as Quebec in 2001, when Lindros sat out a full season, refusing to report. Unfortunately, the deal they subsequently made didn't result in the same kind of windfall.

As an interesting side note, the New York Rangers made an offer for Lindros that would have been even more one-sided, and potentially the worst in NHL history. It was rumoured to include rookies Doug Weight and Tony Amonte, prospect Alexei Kovalev, starting goalie John Vanbiesbrouck, veteran James Patrick, and first-round selections in each of the next three seasons. Yowza.

4. 2000 NHL ENTRY DRAFT:

New York Islanders trade Roberto Luongo and Olli Jokinen to the Florida Panthers for Mark Parrish and Oleg Kvasha.

Mike Milbury's tenure as the GM of the New York Islanders was a rocky one, especially from 1998 through 2001, when three top contenders for this list were made (see trade 6), not to mention some of the era's worst contracts.

Of all Milbury's deals, the worst turned out to be the one in which he dealt away the team's best goalie, 21-year-old fourth-overall selection Roberto Luongo, along with 21-year-old third-overall selection Olli Jokinen (whom he had previously acquired in the sell-off of superstar Ziggy Palffy).

In exchange, Milbury received two older and lesser prospects who would enjoy only occasional flirtations with solid top-six play. Mark Parrish and Oleg Kvasha combined for 473 points from that point forward, while Jokinen by himself has scored 708 points. Luongo was arguably the best goalie of the 2000s, and he currently ranks second among active goalies in shutouts and third in save percentage.

PLAYER	SEASONS	GP	G	A	PTS	GVT
Roberto Luongo*	14	840			2.48	311.2
Olli Jokinen*	14	1,075	301	407	708	96.9
FLORIDA TOTAL	28	1,915	301	407	708	408.1
Mark Parrish	10	568	166	140	306	41.9
Oleg Kvasha	5	337	64	103	167	21.5
NY ISLANDERS TOTAL	15	905	230	243	473	63.4

*Player was still active at the time of writing.

While the Islanders enjoyed some very limited success following these bad deals, they most certainly could have been a multi-time Stanley Cup contender if they had kept Roberto Luongo, Olli Jokinen, Zdeno Chara, Jason Spezza, Todd Bertuzzi, and Bryan McCabe instead.

5. SUMMER 1992:

Chicago Blackhawks trade Dominik Hasek to the Buffalo Sabres for Stéphane Beauregard and 1993 fourth-round pick (Eric Daze).

In 1992, Dominik Hasek was a 27-year-old backup goalie with limited NHL experience, so could Hawks GM Mike Keenan be faulted for swapping him for a late pick and Winnipeg's 24-year-old backup goalie, who had been recently acquired by Sabres GM Gerry Meehan? Yes.

Hasek may have been new to the NHL, but he had established himself in international competition and in Czechoslovakia, where he was named the league's best goalie for five straight seasons. The Dominator proved he was a superstar almost immediately upon arriving in Buffalo, leading the league in save percentage for six straight seasons, winning the Vezina Trophy six times, and almost single-handedly carrying the team to the 1999 Stanley Cup Final.

In fairness, Chicago had no way of knowing that Hasek would ultimately become arguably the best goalie in NHL history, and they did already have Vezina-winning Ed Belfour in nets, but actually choosing Jimmy Waite as his backup over Hasek is inexcusable, and so was considering Stéphane Beauregard as his near-equal in trade value.

PLAYER	SEASONS	GP	G	A	PTS	GVT
Eric Daze	11	601	226	172	398	78.7
Stéphane Beauregard	2	29			4.57	-0.7
CHICAGO TOTAL	13	630	226	172	398	78.0
Dominik Hasek	14	710			2.19	402.4

Fortunately for the Hawks, the fourth-round selection turned out to be solid top-six forward Eric Daze, and they also got a couple of decent seasons out of Christian Ruuttu, who was acquired straight-up for Beauregard three days later.

6. 2001 NHL ENTRY DRAFT:

New York Islanders trade Zdeno Chara, 2001 first-round pick (Jason Spezza), and Bill Muckalt to the Ottawa Senators for Alexei Yashin.

Talented Russian superstar and 1999 Hart finalist Alexei Yashin had worn out his welcome in Ottawa. Demanding a new and higher-paying contract almost every single season and committing one money-related public relations mistake after another, Yashin had essentially become the least popular player to ever don a Senators uniform.

After refusing to report, sitting out the entire 1999–2000 season, and then returning to play out the final year of his deal only to preserve his NHL career, it seemed clear to Ottawa GM Marshall Johnston that even Yashin's tremendous talent couldn't overcome the toxic asset he had become and that the Senators would have to accept whatever pittance was offered in trade.

Fortunately for Johnston, the pittance turned out to be 24-year-old, 6-foot-9 giant defenceman Zdeno Chara, who was just about to break into stardom, and 2001's second overall selection, which was used to nab high-scoring sensation Jason Spezza. The former would go on to compete in six All-Star Games and win the 2009 Norris Trophy, while the latter blossomed into one of the league's best playmakers and currently ranks ninth among active players in career assists per game. Together they helped Ottawa earn four straight 100-point seasons, briefly becoming arguably the league's best team but falling just short of the Stanley Cup in 2007.

PLAYER	SEASONS	GP	G	A	PTS	GVT
Jason Spezza*	12	768	268	481	749	173.6
Zdeno Chara*	13	964	163	346	509	170.0
Bill Muckalt	2	78	5	11	16	3.2
OTTAWA TOTAL	**27**	**1,810**	**436**	**831**	**1,274**	**346.8**
Alexei Yashin	5	346	119	171	290	42.5

*Player was still active at the time of writing.

As for Yashin, he signed a long-term, lucrative deal that no one could have possibly lived up to, and one the Islanders would ultimately choose to buy out—and which continued to use up over $2.2 million of their annual cap space until the 2013–14 season.

7. SUMMER 1979:

Detroit Red Wings trade André St. Laurent, 1980 first-round pick (Larry Murphy), and 1981 first-round pick (Doug Smith) to the Los Angeles Kings for Dale McCourt.

Detroit GM Ted Lindsay's off-season signing of Los Angeles goalie Rogie Vachon kicked off an extended affair that involved lengthy negotiations, court battles, the top trade on this list, the Supreme Court, the NHLPA, the establishment of modern NHL free agency regulations—and Dale McCourt, the first overall selection of the 1977 NHL Amateur Draft.

McCourt was awarded to Los Angeles as compensation for the signing but refused to report. Eventually the two teams, which had been angrily feuding on and off ever since the Kings had signed Detroit's Marcel Dionne five years earlier, managed to settle on a deal to return the talented youngster back to Hockey Town.

Unfortunately for the Red Wings, the entire affair took a psychological toll on the youngster, who never realized his full potential as a game-changing franchise player and ultimately took his talents to the Swiss National League at age 27.

PLAYER	SEASONS	GP	G	A	PTS	GVT
Larry Murphy	21	1,615	287	929	1,216	310.9
Doug Smith	9	535	115	138	253	13.1
André St. Laurent	5	230	42	51	93	5.0
LOS ANGELES TOTAL	**35**	**2,380**	**444**	**1,118**	**1,562**	**329.0**
Dale McCourt	5	377	133	202	335	41.4

Meanwhile, Kings GM George Maguire landed a useful veteran centre in André St. Laurent and used the two draft picks to land another useful centre, Doug Smith, second overall in 1981 and future Hall of Fame defenceman Larry Murphy fourth overall in 1980. If he hadn't gifted Murphy to the Washington Capitals four years later (see trade 18), this one deal might have salvaged the whole Vachon situation.

8. 1979–80 NHL TRADE DEADLINE:

Los Angeles Kings trade 1982 first-round pick (Phil Housley) to the Buffalo Sabres for Jerry Korab.

It's late in the 1979–80 NHL season, running up against the league's first trade deadline, and the Kings still haven't learned their lesson, despite getting repeatedly burnt over the years by trading away draft choices that were ultimately used by other teams on Steve Shutt, Dick Redmond, Ray Bourque, and Larry Robinson.

In this case, GM George Maguire acquired huge 31-year-old puck-moving defenceman Jerry Korab down the stretch. He would play parts of five more seasons, including a strong 52-point campaign in 1980–81.

However, the sixth overall selection he sacrificed was used on Phil Housley, whose 338 career goals and 1,232 points currently rank fourth all-time among NHL defencemen. Housley played 21 seasons, competed in seven All-Star Games, and topped 60 points a dozen times—only Ray Bourque and Paul Coffey managed that feat more often.

PLAYER	SEASONS	GP	G	A	PTS	GVT
Phil Housley	21	1,495	338	894	1,232	307.1
Jerry Korab	6	284	21	99	120	20.3

Sadly, the Kings would repeat this error once again the following year, trading away the first-round selection used on Tom Barrasso in exchange for the final four games of Rick Martin's illustrious career.

9. LATE IN THE 1963–64 SEASON:

Toronto Maple Leafs trade Bob Nevin, Dick Duff, Arnie Brown, Bill Collins, and Rod Seiling to the

> New York Rangers for Andy Bathgate and Don McKenney.

Since Andy Bathgate ultimately scored the Stanley Cup–winning goal that year, it might seem reasonable to give Toronto GM Punch Imlach a pass on one of the league's earliest trade deadline blockbusters. But you have to wonder, could they have won the Cup without the Rangers' captain and future Hall of Famer, and potentially won several more with the tremendous young depth the team sacrificed that day?

Consider what the Leafs gave up and what it could have meant over the long run. The five young players combined for 58 more NHL seasons, including five All-Star Game appearances and one future Hall of Famer in Dick Duff. Toronto could very well have extended its dynasty by another decade.

Instead, it was outgoing Rangers GM Muzz Patrick who helped reverse his team's course. A team that had made the playoffs only six times in the past 22 seasons (in a six-team league) became one of the league's strongest teams for the next decade, and it would begin a successful stretch of qualifying for the postseason in all but four seasons over the next 31 years.

PLAYER	SEASONS	GP	G	A	PTS	GVT
Rod Seiling	15	976	62	267	329	108.4
Bob Nevin	13	878	252	319	571	102.4
Arnie Brown	10	675	44	141	185	54.9
Bill Collins	11	768	157	154	311	33.7
Dick Duff	9	448	109	121	230	29.7
NY RANGERS TOTAL	58	3,745	624	1,002	1,626	329.1
Andy Bathgate	6	350	77	167	254	38.8
Don McKenney	4	130	25	45	70	3.8
TORONTO TOTAL	10	480	102	212	324	42.6

The Leafs may have won the Cup that year, but their new 31-year-old superstar would struggle with knee injuries and get shipped off to Detroit a year later—a few weeks before the Red Wings would also pluck Don McKenney off waivers.

10. 1968 NHL AMATEUR DRAFT:

Los Angeles Kings trade 1969 first-round pick (Dick Redmond) and 1972 first-round pick (Steve Shutt) to the Montreal Canadiens for Gerry Desjardins.

In the first of repeatedly unlucky bloopers, the Kings dealt away future first-round draft picks for urgently needed talent. In contrast, Canadiens GM Sam Pollock would prove to be the master of stealing such selections over the years, even securing Guy Lafleur from the other California-based team two years later (see trade 20).

In Gerry Desjardins, the Kings received a solid young goalie, but they were a struggling expansion team that couldn't take full advantage of him. In fact, Desjardins recorded shutouts in 28% of his wins in Los Angeles. After acquiring Rogie Vachon, Desjardins was dealt away less than two years later, having posted an unfortunate 0.308 winning percentage in 99 games with the Kings.

Years later, Desjardins would finally find his stride in Buffalo, leading that team to the Stanley Cup Final in 1975 and competing in the 1977 All-Star Game. His career was tragically ended at its peak after he was struck in the eye with a puck the following season.

Meanwhile, Montreal invested the fourth overall 1972 selection in future Hall of Famer Steve Shutt, who would go on to become one of the most prolific goal scorers in history, appearing in three All-Star Games, winning five Stanley Cups, and becoming the third player in history to record 60 goals in a single season, in 1976–77.

PLAYER	SEASONS	GP	G	A	PTS	GVT
Steve Shutt	13	930	424	393	817	172.4
Dick Redmond	13	771	133	312	445	120.1
MONTREAL TOTAL	**26**	**1,701**	**557**	**705**	**1,262**	**292.5**
Gerry Desjardins	10	331			3.29	9.3

As for the 1969 selection, it was traded to Minnesota in exchange for not selecting Dick Duff in that year's expansion draft. With it, the North Stars selected Dick

Redmond fifth overall, who scored 445 points in a solid 13-season NHL career.

TRADING DRAFT PICKS IS RISKY

Half of the most lopsided trades so far have involved draft picks, which further reinforces how dangerous that practice can be. For example, in 1983, Montreal traded Robert Picard to the Winnipeg Jets for the 51st overall 1984 entry draft selection, which was ultimately used to select Patrick Roy, arguably the greatest netminder in history.

If trades involving draft picks that fell outside the top 10 were also included, they would dominate this list. It isn't entirely fair to include such trades, however, given that both of the teams involved, not to mention every other club in the league, passed on such players—possibly repeatedly.

At the 1990 entry draft, for another example, Buffalo actually lost a deal by trading *up* in a transaction. They swapped a package that included Phil Housley for Dale Hawerchuk and chose Brad May 14th overall, while the Jets snagged Keith Tkachuk 19th. That same year, the Flames lost a deal with New Jersey when they also traded up to get Trevor Kidd 11th overall, while the Devils selected Martin Brodeur with the 20th overall selection. Those weren't bad deals; they were bad draft picks or, at the very least, unfortunate ones.

Were such deals included, the following five trades would also appear in close succession after the upcoming trade:

1. In 1985, Calgary traded Kent Nilsson and a 1986 third-round selection (Brad Turner) to the Minnesota North Stars for a 1987 second-round selection (Stéphane Matteau) and 1985's 27th overall selection, which they used to draft future Hall of Famer Joe Nieuwendyk.

2. In 1987, Quebec traded Dale Hunter and Clint Malarchuk to the Washington Capitals for Gaétan Duchesne, Alan Haworth, and 1987's 15th overall selection, which they used to draft future Hall of Famer, 12-time All-Star, and 2001 Hart Trophy–winner Joe Sakic.

3. Later in 1987, Buffalo traded Paul Cyr and a 10th-round selection (Eric Fenton) to the New York Rangers for Mike Donnelly and 1988's 89th overall selection, which they used to draft four-time All-Star and top-50 all-time goal scorer Alexander Mogilny.

4. In 1990, as compensation for signing free agent Guy Lafleur, Quebec sent the New York Rangers the 85th overall selection, which they used to draft a three-time All-Star and the top-20 all-time scoring leader among defencemen, Sergei Zubov.

5. In 1997, as compensation for signing free agent Chris Gratton, Tampa Bay received Philadelphia's

next four first-round draft picks, which turned out to be Simon Gagné 22nd overall in 1998, Maxime Ouellet 22nd overall in 1999, Justin Williams 28th overall in 2000, and Tim Gleason 23rd overall in 2001. Thankfully for the Flyers, the Lightning immediately traded all of those picks back in exchange for Mikael Renberg and Karl Dykhuis.

While transactions that involve well-chosen draft picks are outside the scope of this analysis, it's important to be truly mindful of how dramatically the acquisition of even a middling pick can improve a franchise's fortunes.

11. SUMMER 1957:

Detroit Red Wings trade Ted Lindsay and Glenn Hall to the Chicago Black Hawks for Johnny Wilson, Forbes Kennedy, Hank Bassen, and Bill Preston.

Ostensibly as punishment for trying to organize a union in 1957, the Detroit Red Wings sent their superstar "Terrible" Ted Lindsay to the last-place Chicago Black Hawks (as the Blackhawks were known prior to 1986). Having recently reacquired their superstar goalie Terry Sawchuk (for Johnny Bucyk in another horribly lopsided deal), they also included their phenom puck stopper Glenn Hall.

While one would expect a tremendous package in exchange for these two All-Stars, the legendary Red

Wings GM Jack Adams received only a backup goalie, a low-scoring grinder, a long-shot prospect, and only one established NHLer in Johnny Wilson.

The key to making this trade so lopsided was definitely "Mr. Goalie," Glenn Hall, who was arguably the best goalie of the 1960s. He had struggled through a disappointing postseason and didn't agree with Adams's assessment that he would be nothing more than a backup, and he was thus included in this franchise-making gift.

PLAYER	SEASONS	GP	G	A	PTS	GVT
Glenn Hall	14	758			2.57	272.7
Ted Lindsay	4	275	58	93	151	24.2
CHICAGO TOTAL	**18**	**1,033**	**58**	**93**	**151**	**296.9**
Johnny Wilson	5	309	63	76	139	15.7
Forbes Kennedy	10	534	62	95	157	1.5
Hank Bassen	7	123			2.93	0.2
Bill Preston	0	0	0	0	0	0.0
DETROIT TOTAL	**22**	**968**	**125**	**171**	**296**	**17.4**

Ultimately, Hall became one of the most accomplished goaltenders in history, played 14 more seasons, won three Vezina Trophies, set an NHL record with seven First Team All-Star selections, and almost single-handedly turned the notoriously horrible Black Hawks into Stanley Cup champions in 1961. The first-ballot Hall of Famer and grandfather of the butterfly style currently ranks eighth in wins and fourth in shutouts and holds the NHL record for most consecutive games played by a goaltender (502). Not a bad throw-in.

12. 1987–88 TRADE DEADLINE:

> Calgary Flames trade Brett Hull and Steve Bozek to the St. Louis Blues for Rob Ramage and Rick Wamsley.

It's always a gamble to trade away early picks or prospects, especially when it's the obscenely talented son of a hockey legend, like Brett Hull. Though he had set numerous scoring records before going to the NHL, where he had quickly established himself with 50 points in 52 games by the 1987–88 trade deadline, Hull didn't exactly have a reputation as a well-conditioned and determined athlete, nor was he considered someone with any commitment to defensive play.

Having won the Stanley Cup the following season, some people argue that this trade wasn't really the complete disaster for Calgary that it appears to be, but they're wrong. Veteran defender Rob Ramage and backup goalie Rick Wamsley did very little to help the Flames win anything, especially compared to the huge success Brett Hull would become. Thanks to the assistance of coach Brian Sutter and amazing playmaker Adam Oates, Hull was quickly on his way to becoming the third-highest goal scorer of all time. Certainly the Flames should have gotten more than Ramage and Wamsley for someone with that potential.

PLAYER	SEASONS	GP	G	A	PTS	GVT
Brett Hull	18	1,212	717	626	1,343	265.0
Steve Bozek	5	256	54	52	106	14.8
ST. LOUIS TOTAL	**23**	**1,468**	**771**	**678**	**1,449**	**279.8**
Rob Ramage	7	369	31	105	136	21.2
Rick Wamsley	6	122			3.30	8.3
CALGARY TOTAL	**13**	**491**	**31**	**105**	**136**	**29.5**

Interestingly, Oates himself was acquired a year later in another crazy deal, this time for near-retirement veterans Tony McKegney and Bernie Federko. Not a bad year for Blues GM Ron Caron.

13. LATE IN THE 1970–71 SEASON:

Philadelphia Flyers trade Bernie Parent and 1971 second-round pick (Rick Kehoe) to the Toronto Maple Leafs for Mike Walton, Bruce Gamble, and 1971 first-round pick (Pierre Plante).

In an almost fatal miscalculation, the Philadelphia Flyers dealt their starting goalie, Bernie Parent, along with their second-round selection to the Toronto Maple Leafs for Toronto's aging backup Bruce Gamble, a first-round pick, and a top-line centre they needed to complete a separate deal, Mike Walton.

Unfortunately for Toronto, the Maple Leafs squandered their two amazing acquisitions, alienating Parent to the point where he was the first big-name star to depart for the new WHA. Rick Kehoe was an excellent draft choice, but after a couple of solid seasons he was dealt for yet another player who bolted to the WHA, Blaine Stoughton.

Taking some more of the edge off this bad deal for the Flyers, they not only got Parent back almost as easily two years later (see trade 15), both their draft choice and the Walton deal worked out very well. They drafted Pierre Plante, who was traded to the Blues for André Dupont, both of whom served their respective team's blue lines well for at least a half-dozen seasons. For Walton, they acquired two prospects, one of whom turned into a high-scoring two-way forward who would centre their top line for nine seasons, Rick MacLeish.

PLAYER	SEASONS	GP	G	A	PTS	GVT
Bernie Parent	8	363			2.31	204.3
Rick Kehoe	14	906	371	396	797	112.9
TORONTO TOTAL	22	1,269	371	396	797	317.2
Mike Walton	7	331	137	140	277	42.2
Pierre Plante	9	599	125	172	297	26.3
Bruce Gamble	2	35			3.09	2.3
PHILADELPHIA TOTAL	18	965	262	312	574	70.8

Though things ultimately worked out very well for the Flyers, and not nearly as well for the Leafs, this trade could have just as easily turned out as the greatest trade in Toronto's history and Philadelphia's (ahem) second worst.

14. SUMMER 1986:

Vancouver Canucks trade Cam Neely and Glen Wesley to the Boston Bruins for Barry Pederson.

Barry Pederson was on the road to stardom, having scored an amazing 315 points over three seasons by age 22. Despite having lost a step after his shoulder surgery, Vancouver GM Jack Gordon felt Pederson's upside was worth the gamble and gave up both the team's first-round draft choice and rugged 20-year-old prospect Cam Neely to acquire him.

Unfortunately, the 1987 draft was a rather strong one, and that first-round selection was used on great two-way defenceman Glen Wesley, who was selected third overall after Pierre Turgeon and Brendan Shanahan. To make matters worse (for Vancouver), Cam Neely broke out in his very first season in Boston, scoring an amazing 344 goals in his career's remaining 525 games on his way to the Hall of Fame.

PLAYER	SEASONS	GP	G	A	PTS	GVT
Glen Wesley	20	1,457	128	409	537	151.2
Cam Neely	10	525	344	246	590	123.8
BOSTON TOTAL	**30**	**1,982**	**472**	**655**	**1,127**	**275.0**
Barry Pederson	6	322	72	165	237	31.6

As for Pederson, even though he had two perfectly solid seasons in Vancouver, he didn't fulfill his tremendous potential. In the end, it was a gamble that kept the Canucks out of the playoffs and at home watching the Bruins in the 1988 and 1990 Stanley Cup Finals.

15. 1973 NHL AMATEUR DRAFT:

Toronto Maple Leafs trade Bernie Parent and 1973 second-round pick (Larry Goodenough) to the Philadelphia Flyers for 1973 first-round pick (Bob Neely) and Doug Favell.

After one season in the WHA, Bernie Parent wanted back in the NHL but only to Philadelphia (see trade 13). In exchange for regaining his rights, the Flyers sent to Toronto the goalie who had taken Parent's job two years earlier, Doug Favell, and their first-round draft choice for Toronto's second—a deal that certainly didn't look that bad on paper. On the ice, however, the trade turned out much differently.

PLAYER	SEASONS	GP	G	A	PTS	GVT
Bernie Parent	6	298			2.25	182.4
Larry Goodenough	6	242	22	77	99	23.4
PHILADELPHIA TOTAL	**12**	**540**	**22**	**77**	**99**	**205.8**
Doug Favell	6	158			3.70	-10.5
Bob Neely	5	283	39	59	98	-14.6
TORONTO TOTAL	**11**	**441**	**39**	**59**	**98**	**-25.1**

In one of the most dominant goaltending performances in history, Bernie Parent became virtually impenetrable for two seasons, winning the Stanley Cup, the Vezina, and the Conn Smythe both seasons, something he could have just as easily done for Toronto. Briefly derailed by a back injury, Parent was otherwise solid for three more years before an eye injury ended his career at age 33.

16. LATE IN THE 1992–93 SEASON:

> Buffalo Sabres trade Dave Andreychuk, Daren Puppa, and 1993 first-round pick (Kenny Jonsson) to the Toronto Maple Leafs for Grant Fuhr and 1995 fifth-round pick (Kevin Popp).

Players with great reputations tend to be involved in lopsided trades, and while the Toronto Maple Leafs overpaid to get Grant Fuhr in the first place, they more than made up for it a year and a half later when he moved on.

The 1992–93 season was an interesting time in Buffalo. The Sabres had acquired two of the league's best playmakers in Dale Hawerchuk and Pat LaFontaine (see trade 17) and one of the league's best goal scorers in Alexander Mogilny, but they hadn't yet figured out what they had acquired from Chicago in goaltender Dominik Hasek (see trade 5). Therefore, GM Gerry Meehan and coach John Muckler agreed that they could part with their star, 29-year-old forward Dave Andreychuk, in order to upgrade their goaltending situation.

While Fuhr would briefly regain form a few years later in St. Louis, he was nothing more than a mediocre puck stopper at this point in Buffalo, making the deal bad even if it had been Andreychuk for Fuhr straight up. Andreychuk, after all, scored 99 points the following season, played until he was 42 years old, and currently sits 16th in career goals scored.

PLAYER	SEASONS	GP	G	A	PTS	GVT
Dave Andreychuk	13	802	272	262	534	102.1
Kenny Jonsson	10	686	63	204	267	67.9
Daren Puppa	8	214			2.67	52.4
TORONTO TOTAL	31	1,702	335	466	801	232.4
Grant Fuhr	8	350			2.97	-5.8
Kevin Popp	0	0	0	0	0	0.0
BUFFALO TOTAL	8	350			2.97	-5.8

Somehow the Leafs' clever GM Cliff Fletcher managed to convince one of the league's best front offices to throw in the team's backup, Daren Puppa, and its first-round selection, which turned out to be capable top-four defenceman Kenny Jonsson. Though he was selected 15th overall, the deal was already quite lopsided even without that lucky break.

17. WEEKS INTO THE 1991–92 SEASON:

Buffalo Sabres trade Pierre Turgeon, Uwe Krupp, Benoit Hogue, and Dave McLlwain to the New York Islanders for Pat LaFontaine, Randy Hillier, Randy Wood, and 1992 fourth-round pick (Dean Melanson).

New York's 26-year-old superstar Pat LaFontaine, who had just enjoyed his fourth straight season with 40 goals and 40 assists, wanted a new contract that was a little too rich for the Islanders' blood, felt that ownership had broken its promises, and refused to report. Fortunately for the Islanders, GM Bill Torrey found a willing trading partner in Gerry Meehan of the Buffalo Sabres, whose high-scoring 22-year-old phenom Pierre Turgeon was having some frustrations of his own playing second fiddle to the newly acquired Dale Hawerchuk.

At first the deal worked out perfectly for both teams. Over the next two seasons, Turgeon scored 96 goals and 219 points over 152 games, and LaFontaine bagged 99 goals and 241 points in 141 games. However, age and injury collaborated to gradually extinguish LaFontaine's career, while Turgeon continued to shine for a dozen more seasons.

The throw-ins also favoured the Islanders, as Benoit Hogue and Uwe Krupp both enjoyed long and solid careers. Add this to the deal in which the Islanders acquired the first-round selection used on Pat LaFontaine from Colorado for Bob Lorimer and Dave Cameron, and you have the most lucrative trading sequence in the team's history.

PLAYER	SEASONS	GP	G	A	PTS	GVT
Pierre Turgeon	15	972	393	611	1,004	208.5
Benoit Hogue	11	667	177	254	431	70.1
Uwe Krupp	10	402	44	134	178	50.7
Dave McLlwain	6	263	48	59	107	1.6
NY ISLANDERS TOTAL	**42**	**2,304**	**662**	**1,058**	**1,720**	**330.9**
Pat LaFontaine	7	335	181	266	447	80.7
Randy Wood	6	360	81	81	162	24.7
Dean Melanson	1	4	0	0	0	−0.2
Randy Hillier	1	28	0	1	1	−2.1
BUFFALO TOTAL	**15**	**727**	**262**	**348**	**610**	**103.1**

In contrast, the extension of this exchange gets even worse for the Sabres, who dealt away Phil Housley and the first-round selection, which they used on Keith Tkachuk, to acquire the player whose arrival precipitated this situation in the first place, Dale Hawerchuk. Just like with the Rogie Vachon deals, including the trade coming up next, it's fascinating how interconnected these moves were and how one one-sided transaction was built on another.

18. WEEKS INTO THE 1983–84 SEASON:

Los Angeles Kings trade Larry Murphy to the Washington Capitals for Brian Engblom and Ken Houston.

Larry Murphy had a record-setting 76 points as a 19-year-old rookie defenceman in 1980–81 and scored 128 more the following two seasons for the Kings. Over those three seasons, only Paul Coffey, Randy Carlyle, and Doug Wilson outscored him.

Unfortunately for the Kings, GM George Maguire made one mistake after another when dealing with his talented star defenceman. First he represented the team in a seriously botched salary arbitration, then he agreed to trade Murphy rather than rectify the error, and then he failed to get anything more in return than a mediocre 29-year-old defenceman and a winger who was quickly approaching retirement.

PLAYER	SEASONS	GP	G	A	PTS	GVT
Brian Engblom	4	264	10	67	77	11.9
Ken Houston	1	33	8	8	16	1.3
LOS ANGELES TOTAL	**5**	**297**	**18**	**75**	**93**	**13.2**
Larry Murphy	18	1,373	235	774	1,009	240.6

Credit Washington GM David Poile for making the right offer at the right time. Over the next five years the Capitals would place no worse than second in their six-team division, while Los Angeles would finish no better than fourth in its five-team division.

19. 1964 NHL AMATEUR DRAFT:

Boston Bruins trade Ken Dryden and Alex Campbell to the Montreal Canadiens for Guy Allen and Paul Reid.

Ken Dryden was once a Bruin. It was 1964, the second year of the NHL amateur draft, and it wasn't a highly publicized event. Montreal selected Claude Chagnon, Guy Allen, and then Paul Reid, while Boston selected Alex Campbell, Jim Booth, and Ken Dryden. A couple of weeks later, new Montreal GM Sam Pollock swapped two of these top picks with Boston GM Lynn Patrick for two of his, unknowingly making one of the best deals in history.

Interestingly, Reid wasn't even aware he was originally drafted by Montreal and traded for Ken Dryden until 38 years later, always assuming he had just been originally drafted by Boston. Dryden himself didn't learn until the 1970s that he was once a Bruin.

PLAYER	SEASONS	GP	G	A	PTS	GVT
Ken Dryden	8	397			2.24	224.9
Alex Campbell	0	0	0	0	0	0.0
MONTREAL TOTAL	8	397			2.24	224.9
Guy Allen	0	0	0	0	0	0.0
Paul Reid	0	0	0	0	0	0.0
BOSTON TOTAL	0	0	0	0	0	0.0

While none of the other draft choices involved in the trade made the NHL, Ken Dryden went on to win six Stanley Cups and five Vezina Trophies in his seven years as the team's starting goalie, becoming one of the top ten goalies in history by anyone's rankings. Bad luck, Boston.

20. PRIOR TO 1970 EXPANSION AND AMATEUR DRAFTS:

California Golden Seals trade 1971 first-round pick (Guy Lafleur) and François Lacombe to the Montreal Canadiens for 1970 first-round pick (Chris Oddleifson) and Ernie Hicke.

In history's first famous example of how risky it is to trade first-round draft choices, the struggling Oakland Seals swapped prospects and first-round selections (from different years) with the mighty Montreal Canadiens. While their new prospect, Ernie Hicke, and their draft choice, Chris Oddleifson, were both useful second-line forwards throughout most of the 1970s, the pick they gave up was used on one of the game's greatest goal scorers, Guy Lafleur. In an almost hilarious footnote, Montreal GM Sam Pollock later dealt veteran Ralph Backstrom to the Los Angeles Kings to help them pass the Seals the next year, ensuring that the pick he traded for would be first overall.

PLAYER	SEASONS	GP	G	A	PTS	GVT
Guy Lafleur	17	1,126	560	793	1,353	269.8
François Lacombe	2	4	0	1	1	−0.2
MONTREAL TOTAL	**19**	**1,130**	**560**	**794**	**1,354**	**269.6**
Chris Oddleifson	9	524	95	191	286	36.5
Ernie Hicke	8	520	132	140	272	15.2
CALIFORNIA TOTAL	**17**	**1,044**	**227**	**331**	**558**	**51.7**

Imagine if Charlie Finley and the Seals hadn't made that deal, retained that first overall selection, and drafted Guy Lafleur (or Marcel Dionne) for themselves instead. Perhaps the five additional years of postseason absences could have been avoided, preventing the eventual move to Cleveland and subsequent merger with the Minnesota North Stars. Who knows, with a few more good moves, could they have survived to be one of the three California-based franchises around today?

Closing Thoughts

While this is admittedly not the perfect, definitive way to examine trades, this objective perspective is interesting and insightful. What amazes me the most are all the common threads in this list. For example, Rogie Vachon getting signed by the Red Wings kicked off a crazy chain of events. The Kings got Dale McCourt in compensation, but they dealt him back to Detroit for the pick used on Larry Murphy, and Murphy was then traded to the Capitals for practically nothing. Meanwhile, the Kings traded the pick used on Ray Bourque to get a replacement goalie, and then they traded another pick used on Phil Housley to get Jerry Korab from Buffalo to fill in the blue line. Then Buffalo traded Housley along with the pick used on Keith Tkachuk to Winnipeg for Dale Hawerchuk. That gave them the surplus scoring to trade Dave Andreychuk for Grant Fuhr, because they hadn't figured out what they had acquired in Dominik Hasek from Chicago. Also, Hawerchuk's arrival displeased Turgeon, who was dealt to the Islanders for Pat LaFontaine, who was earlier acquired for Bob Lorimer and Dave Cameron. That's over half of this list, and it was all linked to Vachon.

I was also surprised that the infamous Wayne Gretzky deal didn't make the top 20. In fact, it didn't even make the top 100. It was actually surprisingly even, once you add it all up. While the Great One enjoyed a long and successful career, his truly league-shattering seasons were already behind him when he headed south. In return, the Oilers received both a prospect (Martin Gélinas) and a draft choice (Martin Rucinsky) who each enjoyed solid careers that, combined together, aren't too far off what Gretzky accomplished the rest of his career in Los Angeles, St. Louis, and New York.

Of the small difference that still remained between those two packages, some of it was made up by whatever advantage the established player the Oilers received (Jimmy Carson) had over those who went with Gretzky (Marty McSorley and Mike Krushelnyski)—especially since Carson would eventually be used to fleece the Detroit Red Wings out of three top-six forwards.

The trade was still a win for the Kings, no doubt, but not the highway robbery that could have taken place in the desperate situation that occurs when a star player wants to move, especially to a specific place. Indeed, the later deal that sent Gretzky to St. Louis for five players who never had an NHL impact was arguably the more one-sided transaction (had he played out his career there instead of New York).

Of course, that's just my interpretation. These debates will never truly end, but having this objective perspective has hopefully raised some interesting points for the next time they come around.

REFERENCE TABLE: *Top 100*

Using GVT and the previously described objective approach, the following reference table ranks the top 100 most one-sided NHL transactions. For ease of reading, there are bold lines dividing them into groups of 10.

The list won't exactly match the above, because it adds in those that involved potentially flukey draft picks. All players acquired via a draft pick will be denoted in *italics* and the trades that involve such picks that fell outside the top 10 will be lightly shaded.

100 MOST ONE-SIDED TRADES OF ALL TIME*

GVT	TEAM	PLAYERS (GVT)	TEAM	PLAYERS (GVT)
538.2	Boston	*Bourque* (492.4)	Los Angeles	Grahame (−45.8)
397.7	Montreal	*Roy* (433.7)	Winnipeg	Picard (36.0)
370.5	Boston	Esposito (316.6), Hodge (155.2), Stanfield (89.5)	Chicago	Martin (131.5), Marotte (74.6), Norris (−15.3)
367.7	Quebec	Forsberg (206), Duchesne (111.7), Ricci (85.2), *Thibault* (60.5), Hextall (51), Simon (37.9), *Baumgartner* (8.4), Huffman (4.6)	Philadelphia	Lindros (197.6)
344.7	Florida	Luongo (311.2), Jokinen (96.9)	NY Islanders	Parrish (41.9), Kvasha (21.5)
341.1	Montreal	Hextall (49.5), *Robinson* (289.2)	Los Angeles	Duff (−2.4)
340.5	Winnipeg	Housley (178.1), Shannon (26.3), *Tkachuk* (225.6)	Buffalo	Hawerchuk (77.4), *May* (12.1)
324.4	Buffalo	Hasek (402.4)	Chicago	*Daze* (78.7), Beauregard (−0.7)
304.3	Ottawa	*Spezza* (173.6), Chara (170.0), Muckalt (3.2)	NY Islanders	Yashin (42.5)

GVT	TEAM	PLAYERS (GVT)	TEAM	PLAYERS (GVT)
287.6	Los Angeles	St. Laurent (5.0), *Murphy* (310.9), *Smith* (13.1)	Detroit	McCourt (41.4)
286.8	Buffalo	*Housley* (307.1)	Los Angeles	Korab (20.3)
286.5	NY Rangers	Duff (29.7), Nevin (102.4), Brown (54.9), Collins (33.7), Seiling (108.4)	Toronto	Bathgate (38.8), McKenney (3.8)
283.2	Montreal	*Redmond* (120.1), *Shutt* (172.4)	Los Angeles	Desjardins (9.3)
279.5	Chicago	Hall (272.7), Lindsay (24.2)	Detroit	Wilson (15.7), Kennedy (1.5), Bassen (0.2), Preston (0)
266.4	Calgary	*Nieuwendyk* (239.9), *Matteau* (41.7)	Minnesota	Nilsson (15.1), *Turner* (0.1)
263.4	Quebec	Duchesne (29.7), Haworth (7.7), *Sakic* (337.8)	Washington	Hunter (69.5), Malarchuk (42.3)
258.4	Buffalo	Donnelly (35.2), *Mogilny* (225.5)	NY Rangers	Cyr (2.3), *Fenton* (0)
256.0	New Jersey	*Brodeur* (298.0), *Harlock* (–5.8), *Gotziaman* (0)	Calgary	*Kidd* (36.6), *Viitakoski* (–0.4)
250.3	St. Louis	Hull (265), Bozek (14.8)	Calgary	Ramage (21.2), Wamsley (8.3)
246.4	Toronto	Parent (204.3), *Kehoe* (112.9)	Philadelphia	Walton (42.2), *Plante* (26.3), Gamble (2.3)
243.4	Boston	*Wesley* (151.2), Neely (123.8)	Vancouver	Pederson (31.6)
243.4	Tampa Bay	*Gagné* (142.9), *Ouellet* (0.1), *Williams* (118.6), *Gleason* (30.5)	Philadelphia	Gratton (48.7)
239.8	NY Rangers	*Zubov* (244.1)	Quebec	Lafleur (4.3)
235.8	Philadelphia	*Gagné* (142.9), *Ouellet* (0.1), *Williams* (118.6), *Gleason* (30.5)	Tampa Bay	Renberg (34.6), Dykhuis (21.7)
230.9	Philadelphia	Parent (182.4), *Goodenough* (23.4)	Toronto	Favell (–10.5), *Neely* (–14.6)

GVT	TEAM	PLAYERS (GVT)	TEAM	PLAYERS (GVT)
228.2	Toronto	Andreychuk (102.1), *Jonsson* (67.9), Puppa (52.4)	Buffalo	Fuhr (–5.8)
227.8	NY Islanders	Turgeon (208.5), Hogue (70.1), Krupp (50.7), McLlwain (1.6)	Buffalo	LaFontaine (80.7), Wood (24.7), *Melanson* (–0.2), Hillier (–2.1)
227.4	Washington	Murphy (240.6)	Los Angeles	Engblom (11.9), Houston (1.3)
224.9	Montreal	Dryden (224.9), Campbell (0)	Boston	Allen (0), Reid (0)
220.7	Montreal	*Schneider* (215.0)	New Jersey	Maley (–5.7)
219.4	Chicago	*Savard* (219.4)	Quebec	a promise (0)
217.9	Montreal	*Lafleur* (269.8), Lacombe (–0.2)	California	*Oddleifson* (36.5), Hicke (15.2)
215.1	Vancouver	Ronning (131.8), Courtnall (83.1), Momesso (18.7), Dirk (8.2), Loney (1.1)	St. Louis	Quinn (17), Butcher (10.8)
212.2	Edmonton	Semenko (–3.4), *Messier* (340.2)	Minnesota	*Broten* (126.9), *Maxwell* (–2.3)
207.4	NY Islanders	*Luongo* (313.0), Jonsson (65.5), Haggerty (–0.1), Hendrickson (–2.7)	Toronto	Schneider (149.1), Clark (19.9), Smith (–0.7)
206.1	St. Louis	Oates (199.2), MacLean (12.8)	Detroit	Federko (5.6), McKegney (0.3)
201.8	Toronto	Sundin (239.7), Warriner (14.6), *Baumgartner* (10.5), Butcher (2.1)	Quebec	Clark (36.8), Wilson (17.3), Lefebvre (11), *Kealty* (0)
196.5	Los Angeles	Dionne (245.6), Crashley (–1)	Detroit	Maloney (26.6), Harper (21.3), Roberts (0.2)
196.2	Montreal	*LeClair* (198.3), *Popovic* (9.7)	St. Louis	*Nordmark* (11.8), *Bartley* (0)
195.3	Boston	Bucyk (249.1)	Detroit	Sawchuk (53.8)
194.4	Philadelphia	Brind'Amour (203.3), Quinn (16)	St. Louis	Sutter (12.5), Baron (12.4)
192.4	Edmonton	Damphousse (161.4), Richardson (29.5), Thornton (22), Ing (–5.1)	Toronto	Anderson (33), Fuhr (–6.3), Berube (–11.3)

GVT	TEAM	PLAYERS (GVT)	TEAM	PLAYERS (GVT)
190.0	Calgary	Gilmour (186.3), Bozek (11.5), Hunter (9.1), Dark (0)	St. Louis	Bullard (18.7), Corkery (0), Coxe (−1.8)
183.1	Edmonton	Murphy (93.5), Graves (85), Klima (46.2), Sharples (0)	Detroit	Carson (43.9), Layzell (0), McClelland (−2.3)
181.2	New Jersey	*Niedermayer* (213)	Toronto	Kurvers (31.8)
181.2	St. Louis	Conroy (113.5), Turgeon (106.2), Fitzpatrick (2.3)	Montreal	Corson (31.3), Baron (9.5), *Razin* (0)
179.0	Ottawa	*Demitra* (178.9)	Toronto	Miller (−0.1)
178.6	Vancouver	McCabe (115), Bertuzzi (100), *Ruutu* (18.8)	NY Islanders	Linden (55.2)
172.4	Edmonton	*Anderson* (172.4)	Minnesota	a promise (0)
171.4	Buffalo	*Miller* (141.1), Warrener (33.9)	Florida	Wilson (3.6)
168.4	Calgary	Iginla (272.5), Millen (4.6)	Dallas	Nieuwendyk (108.7)
168.3	Vancouver	*D. Sedin* (191.2), *Birbraer* (0)	Atlanta	*Stefan* (22.9)
167.6	Pittsburgh	Zubov (197.2), Nedved (102.7)	NY Rangers	Robitaille (113.4), Samuelsson (18.9)
166.9	Detroit	Mahovlich (122), Unger (113), Stemkowski (85)	Toronto	Ullman (98), Henderson (66.1)
166.4	NY Islanders	*LaFontaine* (174.8)	Colorado	Lorimer (8.4), Cameron (0)
162.5	Tampa Bay	McCarthy (1.3), *Richards* (163.4), *Rich* (0)	Calgary	Wiemer (2.2)
160.9	Anaheim	Selanne (244.4), Chouinard (15.7), *Staal* (0)	Winnipeg	Tverdovsky (68.4), Kilger (30.8), *Lundstrom* (0)
159.7	St. Louis	Demitra (160.9)	Ottawa	Olsson (1.2)
156.8	Washington	Ridley (118.5), Miller (85.5), Crawford (−0.5)	NY Rangers	Carpenter (46.7), *Prosofsky* (0)
155.1	San Jose	Ozolinsh (158.9), Nazarov (−3.8)	St. Louis	a promise (0)
153.0	Chicago	Chelios (208.3), Pomichter (0)	Montreal	Savard (55.3)

GVT	TEAM	PLAYERS (GVT)	TEAM	PLAYERS (GVT)
152.0	Boston	Rask (123.3)	Toronto	Raycroft (-28.7)
151.6	Vancouver	Naslund (151.4)	Pittsburgh	Stojanov (-0.2)
151.6	Philadelphia	Crossman (62.6), *Mellanby* (112.1)	Chicago	Wilson (23.1)
151.2	Boston	Middleton (156.5)	NY Rangers	Hodge (5.3)
148.9	NY Rangers	Messier (157), Beukeboom (29.5), Shaw (20.1)	Edmonton	Nicholls (54.5), Rice (9.4), DeBrusk (-6.2)
144.4	Washington	Jagr (138.8), Kucera (4.8)	Pittsburgh	Beech (0.7), Lupaschuk (-0.5), Sivek (-1)
140.0	New Jersey	*Parise* (136.3)	Edmonton	*Pouliot* (2.8), *Jacques* (-6.5)
139.8	Minnesota	Murphy (177.1), Gartner (94.8)	Washington	Ciccarelli (91.3), Rouse (40.8)
139.0	Buffalo	*Chisholm* (0), *Barrasso* (139.6)	Los Angeles	Martin (0.6)
138.2	Philadelphia	Timonen (72.6), Hartnell (72.1)	Nashville	*Blum* (6.5)
136.9	Philadelphia	LeClair (175.7), Desjardins (143.4), Dionne (0.7)	Montreal	Recchi (182.9), *Hohenberger* (0)
135.0	Toronto	Gilmour (120.1), Macoun (26.1), Manderville (15.7), Nattress (7.9), Wamsley (-5.1)	Calgary	Petit (19.1), Godynyuk (8.4), Leeman (7.1), Reese (-0.5), Berube (-4.4)
135.0	Calgary	*Konroyd* (37.3), *Vernon* (93.0)	Boston	Craig (-4.7)
134.6	Colorado	*Tanguay* (166.0), Donovan (15.1)	San Jose	Ricci (47.0), *Kristek* (-0.5)
134.5	Pittsburgh	Hossa (119.0), Dupuis (51.6)	Atlanta	Armstrong (20.0), Christensen (16.1), Esposito (0), *Leveille* (0)
133.8	Boston	*MacLeish* (140.2)	Philadelphia	Paiement (6.4)
133.0	San Jose	Selanne (141.5)	Anaheim	Friesen (23.8), *Polak* (-0.4), Shields (-14.9)
132.5	Vancouver	Luongo (172.4), Krajicek (4.2), *Shirokov* (0.6)	Florida	Bertuzzi (24.5), Allen (27.4), Auld (-7.2)
131.9	St. Louis	*Liut* (140.1), Evans (0.1)	Kansas City	Giroux (8.3)

GVT	TEAM	PLAYERS (GVT)	TEAM	PLAYERS (GVT)
131.6	Nashville	Timonen (129.2), Vopat (2.4)	Los Angeles	a promise (0)
129.7	Tampa Bay	Boyle (129.7)	Florida	Tuma (0)
128.9	Anaheim	*Perry* (131.6)	Dallas	*Polak* (−0.4), *Crombeen* (3.1)
128.4	Buffalo	Brière (99.9), *Sekera* (43.6)	Phoenix	Gratton (15.0), *Reddox* (0.1)
127.0	Philadelphia	*Prospal* (127.2)	Toronto	Bullard (0.2)
125.5	Dallas	Zubov (180.8)	Pittsburgh	Hatcher (55.3)
124.5	Buffalo	Satan (124.4)	Edmonton	Millar (0), Moore (−0.1)
124.2	Hartford	*Pronger* (235.7)	San Jose	Makarov (16.9), *Kozlov* (82.5), *Kroupa* (−0.4), *Peltonen* (12.5)
123.4	Montreal	Mahovlich (153.8), Crashley (2.9)	Detroit	Monahan (33.3)
122.5	Anaheim	Giguère (130.4)	Calgary	Pettinger (7.9)
122.4	Philadelphia	*Linseman* (122.4)	NY Rangers	Coach Shero (0)
120.8	Philadelphia	*Pavelski* (117.3), Harlock (0), *Redenbach* (0)	Atlanta	Francis Lessard (−3.5)
120.5	Boston	Sawchuk (70.9), Stasiuk (67.5), Bonin (48), Davis (−0.8)	Detroit	Godfrey (44.8), Chevrefils (18.9), Sandford (1.9), Boisvert (0.1), Corcoran (−0.6)
120.0	Montreal	*Carbonneau* (120.0)	St. Louis	Roberts (0)
118.8	New Jersey	*Gomez* (123.4)	Dallas	*Erskine* (6.2), *Bouck* (−1.6)
118.8	Los Angeles	*Visnovsky* (136.5)	Ottawa	*Zanon* (17.7), *Potulny* (0)
117.3	San Jose	*Pavelski* (117.3)	Philadelphia	*Scurko* (0)
116.7	Edmonton	Joseph (86.8), Grier (61.3)	St. Louis	*Reasoner* (31.4), *Zultek* (0)
116.3	Chicago	Amonte (158.4), Oates (0)	NY Rangers	Matteau (25.9), Noonan (16.2)
115.5	Boston	*Leach* (110.8)	Los Angeles	Krake (−4.7)

[* Raw player GVT data for these calculations obtained from Hockey Prospectus (blog), http://www.hockeyprospectus.com. http://www.hockeyprospectus.com/gvt-goals-versus-threshold/.]

QUESTIONS AND ANSWERS

by ROB VOLLMAN

There is an endless supply of fascinating questions upon which hockey analytics can shed some light, but not all of them warrant an entire, dedicated chapter of their own. That's why we use this final chapter to cover topics that can be answered briefly or to provide additional insight in specific areas.

In the past, we've used this section to tackle interesting questions, to introduce or expand on new statistics, to cover new developments on material included in previous editions, to comment on current events in the field of hockey analytics, or to expand on concepts covered earlier in the book. This time, we're taking advantage of these pages to establish the value of home-ice advantage and to cover a little bit of the history of the hockey analytics field.

How Great Is the Home-Ice Advantage?

In the NHL, the home team wins roughly 55% of the time. That may sound like a huge advantage, but it's actually the smallest among all major North American sports.

Where does this advantage come from? The absence of travel, getting to train and play in familiar surroundings, and hearing the cheers of a friendly crowd all provide intangible boosts that have yet to be measured. On the other hand, numerous studies have shown that the combined effects of built-in rules advantages on faceoffs and with the last line change, along with favourable officiating, can explain most, if not all, of the home team's advantage. Let's focus in on those latter, measurable advantages.

1. OFFICIATING

Based on at least half a dozen studies over the past decade, the conventional wisdom that the home team gets a few extra calls has a great deal of basis in fact. In essence, the question is not whether this advantage exists but how great it is.

In the first attempt to quantify the advantage, Chris Boersma found that home teams are tagged on only 48.4% of penalties that require the referee's personal discretion and also observed that home teams enjoy a lot more five-on-three situations.*

[* Chris Boersma, "Home Team Advantage," *Hockey Numbers* (blog), October 31, 2006, http://hockeynumbers.blogspot.ca/2006/10/home-team-advantage-i-do-think-that.html.]

A few years later, noted statistician Gabriel Desjardins found that the visiting team averages 0.83 more penalty minutes per game. He was also the first to break down the data to identify which individual officials had more success ignoring the roars of the crowd, but without enough data to draw any firm conclusions.*

[* Gabriel Desjardins, "Advanced Metrics: The Homer Referee Bias," *Arctic Ice Hockey* (blog), November 30, 2009, http://www.arcticicehockey.com/2009/11/30/1173912/referee-homer-bias.]

The topic got mainstream attention in 2009, when Vancouver researcher Will Lockwood published a study in the *Edmonton Journal* that found that the home team enjoyed 11.5% more power plays than its opponents.* That result surprised the referees, coaches, and players who were interviewed, but it has been backed up repeatedly since then.

[* Curtis Stock, "Study Reveals Referees' Home Bias," *Edmonton Journal,* March 2, 2009, http://www.edmontonjournal.com/sports/hockey/edmontonoilers/Study+reveals+referees+home+bias/1343770/story.html.]

- Using 2008–09 data and accounting for score effects, Jack Brimberg and William J. Hurley found that the home team enjoys 52.5% of a game's power plays, which works out to 10.5% more than the visiting team.*

[* Jack Brimberg and William J. Hurley, "Are National Hockey League Referees Markov?," OR Insight 22 (December 2009): 234–243, http://www.palgrave-journals.com/ori/journal/v22/n4/full/ori200912a.html.]

- In the controversial book *Scorecasting,* it was contended that home teams actually enjoy up to 20% more power plays than the visiting team.[35]
- More recently, Wesley Chu found that teams take 48% of their penalties at home, and he also found that, at the end of the season, teams that take fewer penalties at home outnumber those who are more lightly penalized on the road by almost a five-to-one margin.*

[* Wesley Chu, "Home Ice Advantage," *Hockey Metrics* (blog), July 10, 2013, http://hockeymetrics.net/home-ice-advantage/.]

While a notable factor, home-ice advantage isn't entirely about favourable officiating. After all, teams

---

35 Tobias J. Moskowitz and L. Jon Wertheim, Scorecasting: The Hidden Influences Behind How Sports Are Played and Games Are Won (New York: Three Rivers Press, 2012), 157.

also score more at even strength, not just with the man advantage. The higher winning percentage at home is a consequence of several factors, some of which are possibly even greater than the unintentional influence of the officials.

2. FACEOFFS

As covered in more detail in the chapter on faceoffs, the home team has a built-in rules advantage that can be particularly valuable in critical draws in the defensive zone and/or on special teams. Specifically, rule 76.4 in the official NHL rule book states that "the sticks of both players facing-off shall have the blade on the ice, within the designated white area. The visiting player shall place his stick within the designated white area first followed immediately by the home player."*

[* National Hockey League, *National Hockey League Official Rules 2014–15,* NHL, http://www.nhl.com/nhl/en/v3/ext/rules/2014-2015-rulebook.pdf.]

Getting to place their stick down last allows home team centres to win up to 52% of all faceoffs. That may appear to be a small advantage, but it's enough for almost all teams to win more faceoffs at home than on the road and has proven to be the decisive edge on more than one occasion.

As of the 2015–16 season, it is now the team in the offensive zone that enjoys this advantage, not the home team.

3. THE LAST CHANGE

The most important built-in rules advantage comes courtesy of rule 82.1, which states that "following the stoppage of play, the visiting team shall promptly place a line-up on the ice ready for play and no substitution shall be made from that time until play has been resumed. The home team may then make any desired substitution, except in cases following an icing, which does not result in the delay of the game."*

[* National Hockey League, *National Hockey League Official Rules 2014–15,* NHL, http://www.nhl.com/nhl/en/v3/ext/rules/2014-2015-rulebook.pdf.]

This rule allows the home team the final line change after almost every whistle in a game. While the total impact of this advantage is hard to measure and varies from team to team and coach to coach, it's the main reason why the home team scores more at even strength. In my view, this is the largest component in the home-ice advantage.

I remember an early conversation I had with a prominent front office personality about how comparing a player usage chart of home games versus one based on road games was a key tool in his player and coaching evaluations. On the road, teams are at the mercy of the opposing coach's line matchups, while home games more accurately reflect a coach's desired player usage.

That's also why certain players get a lot more ice time at home, where their coaches can properly protect them. Consider Keith Yandle of the New York Rangers. Acquired from Arizona at the trade deadline, the 28-year-old veteran is a great puck-moving defenceman but not the type of player that an aggressive line-matching coach like Alain Vigneault wants on the ice against top talent. That's why Yandle typically enjoys up to 20 minutes of even-strength ice time at home but closer to 18 minutes on the road, where he is harder to shelter. That extra ice time is yet another reason why Yandle, and the Rangers themselves, would be more effective at home.

4. OTHER FACTORS

The NHL rulebook has several other interesting advantages built in, but none that statistically contributes to the home team's success. Specifically, the home team

- selects which end of the ice on which to play;
- heads to the dressing room first;
- supplies the pucks;

- has more advantageous faceoff locations when the appropriate location is unclear;
- may request the time of the last commercial break in the first and second periods;
- decides who will shoot first in a shootout, but this option hasn't historically proven to be particularly helpful.

Additionally, Boersma found that the home team is less likely to be playing on back-to-back nights, and they consequently have 0.2 more days of rest, on average. Furthermore, home teams are more likely to face a backup goalie, especially on the tail end of a back-to-back.* Chu, however, found that extra travel doesn't appear to affect a game's results.

[* Chris Boersma, "Home Team Advantage," *Hockey Numbers* (blog), October 31, 2006, http://hockeynumbers.blogspot.ca/2006/10/home-team-advantage-i-do-think-that.html.]

While each of these various factors seem small by themselves, the home-ice advantage becomes quite significant when it's added up. Teams earn almost 55% of their points at home, which is why teams are over six times more likely to post an equal or better record in their own rink.

What's the History of the Hockey Analytics Community?

When it comes to the use of statistics in sports, baseball has much of its story written, some of which has even been immortalized in the *Moneyball* film starring Brad Pitt. How about hockey?

This is a topic that could easily fill an entire chapter, or even a complete book. And if it ever does become a movie, Dwayne "the Rock" Johnson should obviously be cast to play me.

In all seriousness, as the popularity of hockey analytics began taking off, there has been more and more interest in learning about its roots, both within the NHL and in our online community. Indeed, this was the very topic on which I was asked to present at the grassroots hockey analytics conferences that were organized in Ottawa on February 7, 2015, and in Washington on April 11, 2015. While I am by no means the gatekeeper of our history, I have been engaged in this scene long enough to at least provide a brief, high-level overview.

The story actually begins in the early 1940s, when Dick Irvin Sr. was hired by the Montreal Canadiens and first met local tie salesman Allan Roth. Much like today's bloggers, Roth was an outsider who viewed sports analytics primarily as a hobby. Working with Irvin, he invented an early version of plus/minus, the first hockey statistic that went beyond the basics like games played, goals, assists, and wins.*

[* David Staples, "Just How Horse [* xxxx] Is the NHL's Official Plus/Minus Stat?," *Edmonton Journal,* May 13, 2013, http://blogs.edmontonjournal.com/2013/05/13/just-how-horse-shit-is-the-nhls-official-plusminus-stat/.]

Roth also kept statistics for the local minor league baseball team, the Montreal Royals. In 1947 he was hired by Branch Rickey of the Brooklyn Dodgers, becoming that sport's first full-time statistician.

Fortunately for Irvin, Roth's innovations, including the plus/minus statistic, remained a trade secret for the Canadiens for roughly two decades, before they finally caught on with other teams and were ultimately introduced league-wide for the 1967–68 season. Since this statistic also requires breaking down a team's scoring by manpower situation, the league had rolled out that data as well, in 1963–64. The first analytics hobbyist had made his mark.

The next inspiration came with the 1972 Summit Series tournament between Canada and the Soviet Union. While Soviet coach Anatoli Tarasov was well-known for his training techniques, he was also an early pioneer in hockey analytics. Judging from his 1969 book *Road to Olympus,* Tarasov was tracking zone exits and entries in much the same fashion as today's (previously) independent analysts, like Eric Tulsky and Corey Sznajder. There is even some evidence that Tarasov was tracking attempted shots, which is known today as Corsi and was introduced as part of the NHL.com's enhanced stats package in 2015.

In both his training methods and his approach to hockey analytics, Tarasov's inspiration was Lloyd Percival's *Hockey Handbook,* which was first published in 1951 and further popularized when re-published two years after the Summit Series.

Percival's statistical efforts were closely associated with his training methods and primarily tracked physical factors such as skating speed, both with and without the puck, shot speed, and goalie reaction time. Much like today's independent bloggers Ryan Stimson and Emmanuel Perry, Percival (and Tarasov) also tracked passes.

In 1977, the NHL was introduced to its next analytics pioneer, "Captain Video" Roger Neilson. As his nickname suggests, Toronto's new coach watched hours of videotape every day, counting virtually every event that could be recorded on camera. Most notably, he formally introduced the notion of scoring chances, which he integrated into a new version of plus/minus. Neilson was so aggressive in seeking out advantages anywhere and everywhere he could that there are actually a handful of rules, which remain in effect today, that were specifically designed to prevent some of his more notorious exploits.

The 1970s also saw the debut of the World Hockey Association (WHA), the NHL's greatest professional hockey rival since the old Pacific Coast Hockey Association (PCHA). The WHA opened its doors to new people and new ideas, consequently introducing a number of new statistics, most notably the goalie's save percentage. Despite counting shots and shooting percentage as far back as its 1967 expansion, the NHL didn't introduce save percentage until the 1983–84 season.

Many of the older members of today's hockey analytics community, including the three authors of this book, got interested in the field shortly after the introduction of save percentage. The first (and, until recently, only) hockey analytics book came out in 1987 courtesy of Jeff Klein and Karl-Eric Reif, the *Hockey Compendium.* The two pioneers would go on to contribute to *Total Hockey* and produce another edition of the *Hockey Compendium* years later.

Unlike today's analysts, Klein and Reif were working with virtually no raw data beyond goals, assists, and plus/minus, but they still collaborated on some groundbreaking concepts that would not be improved on for a decade or two. If you have enjoyed this book, I highly recommend finding an old copy on eBay to add to your collection.

The first wave of new information finally came to the NHL for the 1997–98 season, in the form of real-time scoring statistics (RTSS). While some of the new numbers weren't particularly enlightening, like the hits and blocked shots Iain covered in earlier chapters, the recording of player ice time was huge. Knowing who was being used the most at even strength, on the penalty kill, and on the power play was a huge advance, and even statistics as simple as points per minute were a great leap forward. Thinking back, it was a tremendous relief to finally put our ice-time estimates aside, as clever as they may have been at the time.

Shortly after the introduction of RTSS, FOX Sports brought in the much-maligned puck-tracker system. While unpopular from a viewing experience perspective, the data it produced was invaluable. It was our first actual possession and zone-time data, and it proved useful to confirm the close relationship between attempted shots (aka Corsi) and actual possession and zone times. As of the end of the 2014–15 season, we are still waiting for such data to be re-introduced by the NHL or its broadcasting partners.

Around the turn of the century, hockey analytics first began to appear on the Internet. Most notably, this included Iain's *Puckerings* blog and a short-lived column for *Sports Illustrated* penned by Marc Foster called "By the Numbers." The two early pioneers almost launched Hockey Prospectus at that same time, which would have been an early boon.

This is also where I made my formal entrance into the field, enjoying a publishing credit as a co-author of a paper Iain wrote for the *Hockey Research Journal* in 2001. Our article was about fixing the plus/minus statistic by taking the same approach introduced by Klein and Reif but incorporating the new, actual ice-time figures as well as a few other adjustments of our own.

Back then, it wasn't unusual for those getting into hockey analytics to make their debut either by fixing plus/minus or by developing a catch-all statistic. My own contribution to the former was to adjust it for special teams play, since power-play goals aren't counted in the raw statistic but short-handed goals are, which boosts the plus/minus of those who work exclusively in the latter situation at the expense of those who primarily work with the man advantage.

Since we wouldn't have exact totals of goals scored and against while a player is on the ice until after the 2005 lockout, most of our early work involved estimates. Today, estimates are sometimes sneered upon by people new to the hockey analytics game, but without the actual data, that's all we could use back then. In fact, most of today's non-traditional data is still unavailable in other hockey leagues, so analysts like Josh Weissbock take the best of this old work to produce modern-day statistics for leagues like the Canadian juniors.

The movement really started to grow shortly before, during, and after the 2005 NHL lockout. In 2004, Iain started a discussion forum called the Hockey Analysis

Group (HAG), which is where many of us met for the first time. This includes our co-author Tom Awad and other early pioneers, like Alan Ryder, Tom Tango, and Gabriel Desjardins (to name just a few). Many of them were already starting to consult with NHL teams, although without any real publicity.

It wasn't long before Iain's blog got some company, including most notably Alan Ryder's *Hockey Analytics* in 2004 and David Johnson's *Hockey Analysis* in 2005, which is the longest continually active hockey analytics blog today (to my knowledge). Around 2007, Johnson and *Behind the Net'* s Desjardins also added the first two online databases of new, non-traditional hockey statistics to their websites, both of which remain available and free of charge today and which the NHL themselves are working to replicate.

Beyond those of us on HAG, the other primary source of hockey analytics experts was the Oilogosphere, which really started to take form after the lockout. Most of the shot-based statistics and analysis were innovated and/or popularized by this group, which featured Timothy Barnes (aka Vic Ferrari), Sunny Mehta, Tyler Dellow, James Mirtle, Allan Mitchell, and many others who are probably less recognizable than the statistics that were named after them, like Matt Fenwick and Brian "PDO" King. Of course, there were also those who fell outside either umbrella, like Johnson and Tore Purdy.

The next big step forward occurred in 2009, when Hockey Prospectus was finally launched, primarily with analysts from HAG, at roughly the same time as SB Nation, which eventually brought in a whole new generation of great minds, most notably Eric Tulsky. While it was previously possible to keep track of everyone involved in hockey analytics and every single development from the various websites and discussion forums, those days were coming to an end.

Why didn't an NHL team hire some of these key innovators back then, to essentially gain a monopoly on hockey analytics? Imagine the advantage a team would have with exclusive ownership of these advances, much as Dick Irvin enjoyed with the Montreal Canadiens through the 1940s and 50s.

Actually, this was attempted. Some time around the 2010–11 season, the Phoenix Coyotes hired Desjardins, Barnes, Dellow, Awad, and Mehta as a group of consultants known as "the Quants."[36] Unfortunately (for the Coyotes), this group was disbanded after the departure of prospective owner Matthew Hulsizer.

Even if the Coyotes had held on to these great minds, many of their advances had already been published, and they now represented only a fraction of the available pool of expertise. Even in 2011, several more

---

36 Bob McKenzie, Hockey Confidential: Inside Stories from People Inside the Game (Toronto: Harper Collins Canada, 2014), 89.

"dream teams" could have been assembled before the pool dried up. I distinctly remember the joy of constantly discovering brand-new websites, with brand-new analysts and ideas and with increasing regularity.

I knew it was only a matter of time before something like 2014's summer of analytics would occur, which saw most of the names I just mentioned get hired into NHL front offices or into mainstream media outlets like TSN and Sportsnet. Even the NHL itself rolled out a brand-new slate of stats, most of which were innovated 50 years ago by the likes of Percival and Tarasov and formalized by their modern-day successors. The genie was definitely out of the bottle.

Again, I'm not ideally suited to be history's gatekeeper, and I didn't intend this to be a comprehensive account of who we are and where we came from, but I can tell you that it has been a fun ride for all of us who have been inspired by legends like Roth, Percival, Tarasov, and Neilson. We can only hope that we have continued their work and motivated a new generation to take it forward from here.

CONCLUSION

Once again, this is the only conclusion you'll find in one of our books. Analytics can certainly shed light on an issue, but the great debates never reach a definitive conclusion.

Thanks to our illustrator Joshua Smith, special advisors Alan Ryder, Craig Tabita, and Phil Myrland, and to everyone who has enjoyed our work, supported us, and helped spread the word.

Remember that all the raw data for the charts and tables can be found on the Hockey Abstract website. Our Twitter handles and emails can be found in the author biographies. Please don't be shy about reaching out to us.

Let the discussions continue.

GLOSSARY

This is only a quick-and-dirty reference of analytics terms used in this book.

3-1-1 RULE

Three goals get a team about one point in the standings and cost about $1 million.

ADVANCED STATISTICS

All statistics beyond those recorded in newspaper box scores as well as non-traditional statistics.

AGE CURVE

A calculation of how players' performances change as they age.

BEHIND THE NET

A hockey statistics data website and blog hosted by Gabriel Desjardins.

BILL JAMES

Author of the *Baseball Abstract* and a pioneer of the baseball analytics revolution.

CATCH-ALL STATISTIC

A statistic, like GVT, that captures a player's entire value in a single number.

CLOSE GAME

A game within a single goal in the first two periods or tied in the third period or overtime.

CORRELATION

A statistical term that calculates the extent to which two variables are related.

CORSI/SAT

A differential based on all attempted shots, including those blocked or that missed the net.

DEAD MONEY

Cap space allocated to bought-out contracts, prior bonuses, and retained salary.

DEFENCE-INDEPENDENT GOALIE RATING (DIGR)

Save percentage adjusted based on the average quality of shots faced.

DELTA, DELTASOT

A shot-based, situation-neutral statistic developed by Tom Awad in 2010.

EVEN-STRENGTH BLOCKED-SHOT PERCENTAGE (EBS%)

Percentage of all opposing shot attempts blocked by an individual player.

EXPECTED SAVE PERCENTAGE (XSV%)

A league-average save percentage, given the location and quality of the shots a goalie faced.

FENWICK/USAT

A differential based on all attempted shots, including those that missed the net but not including blocked shots (unlike Corsi).

GOALS VERSUS SALARY (GVS)

The number of goals scored or prevented relative to a player's cap hit.

GOALS VERSUS THRESHOLD (GVT)

The value of a player's contributions in terms of goals relative to a replacement-level player.

HOCKEY COMPENDIUM

Written by Jeff Klein and Karl-Eric Reif, the first (and, until the first edition of *Hockey Abstract,* only) book of this kind.

HOCKEY PROSPECTUS

A website and blog hosted by Timo Seppa that features statistical hockey analysis.

HOME-PLATE SAVE PERCENTAGE

A goalie's save percentage in shots taken in a dangerous zone in front of the net.

LEAGUE-QUALITY FACTORS

An estimate of the relative difficulty of scoring and preventing goals in a given league.

LOCATION-ADJUSTED EXPECTED-GOALS PERCENTAGE (LAEGP)

League-average save percentage based on the exact location of shots taken.

MONEYBALL

Michael Lewis's book documenting the adoption of analytics by Oakland A's GM Billy Beane.

NET SHOTS POST FACEOFF (NSPF)

Measuring faceoff ability based on post-draw shot differential.

ONE-STAT ARGUMENT

The belief that goaltenders should be evaluated using even-strength save percentage only.

PDO (OR SPSV%)

On-ice shooting percentage plus on-ice save percentage.

PERSISTENCE

A statistical term that measures the extent to which a variable is consistent over time.

PLAYER USAGE CHART (PUC)

A graphical depiction of how a player is being used.

PLUS/MINUS

Long-time NHL statistic representing a player's on-ice goal differential.

POSSESSION BASED

A hockey system aimed at keeping possession of the puck and denying such possession to the opponents.

PROJECTINATOR

A system that uses a player's junior data to predict the mean of his career average GVT to the age of 27 and the average GVT of his two best seasons.

PROSPECTIVE SCOUTING/ANALYSIS

Using a player's past performance to predict his future success.

PYTHAGOREAN EXPECTATION

A formula that predicts wins and losses based on goals for and against.

QUALITY OF COMPETITION (QOC)

A high-level and Corsi-based measure of the average level of competition a player faces.

QUALITY START (QS)

A measurement of whether a goalie played well enough for his team to win.

RECORDING BIAS (OR SCOREKEEPER BIAS)

The bias introduced by the different ways scorekeepers judge game events, like shots.

REGRESSION

The tendency for observed outcomes that include a component of random variation to approach the mean over time.

RELATIVE

Expressing a statistic relative to a player's team and their results without the given player.

REPLACEMENT-LEVEL PLAYER

A general term for a player whose contributions are similar to that of an AHL call-up.

RETROSPECTIVE ANALYSIS/SCOUTING

Backward-focused analysis that examines or explains past success.

SAMPLE SIZE

A statistical term that refers to the amount of data on which an analysis is based.

SCORE EFFECTS

The skewing effect teams either sitting on or chasing leads can have on statistics, most notably shots.

SELECTIVE SAMPLING (OR SELECTION BIAS)

The bias introduced when the manner in which players are selected for an analysis affects the outcome.

SHOT QUALITY

The combined impact of the timing, type, circumstances, and location of a shot, sometimes together with the shooter's skill.

SURVIVORSHIP BIAS

An analysis that is skewed by including only those players strong enough to qualify.

WINS ABOVE REPLACEMENT (WAR)

The value of a player's contributions in terms of wins relative to a replacement-level player.

WOWY (WITH OR WITHOUT YOU)

An analysis comparing how a given player performs with and without a particular linemate.

ZONE STARTS (ZS)

The percentage of all non-neutral shifts a player started in the offensive zone (not counting on-the-fly line changes).

ABOUT THE AUTHORS

ROB VOLLMAN
Twitter: @robvollmanNHL
Email: vollman@hockeyabstract.com
Website: http://www.hockeyabstract.com

Best known for player usage charts and his record-breaking ESPN Insider contributions, Rob Vollman was first published in the fall 2001 issue of the *Hockey Research Journal* and has since co-authored 10 books in the *Hockey Abstrac*t, *Hockey Prospectus,* and *McKeen's* magazine series.

While modern advanced statistical hockey analysis stands on a mountain of complexity, Rob's work is best known for being expressed in clear, focused, and applicable terms, often presented in a humourous and entertaining way. Whether you're arguing about the worst trades in history or which team improved most in the off-season, Rob's objective approach will add clear, cold facts to the discussion in a style that is undeniably engaging—and convincing.

Rob's most popular innovations include player usage charts, quality starts and home-plate save percentage for goaltenders, goals versus salary (GVS) to measure a player's cap value, his history-based projection systems, the setup passes statistic, and advances in the field of NHL translations and league equivalencies

(NHLe) to understand how well players coming from other leagues will perform.

Rob has been a member of the Professional Hockey Writers Association (PHWA) since 2014, and his analyses can be found regularly at *Hockey Prospectus* and ESPN Insider and have been featured in the *Hockey News,* the *Globe and Mail,* the *Washington Post, Forbes,* and *Rolling Stone.* Rob has made 200 appearances on 40 different radio programs, TV shows, and podcasts in 18 NHL cities, including most notably *Hockey Night in Canada* radio, Sportsnet's *Hockey Central,* TSN's *That's Hockey,* ESPN's *SportsCenter* and *Hockey Today,* CBC Radio, and Wharton Business Radio.

IAIN FYFFE
Twitter: @IainFyffe
Email: hockeyhistorysis@gmail.com
Website: http://hockeyhistorysis.blogspot.com

Iain Fyffe has been doing hockey analytics since before there was really such a thing as hockey analytics. He started the Puckerings website in 2001 to publish his work. Later known as Hockeythink, it was the first dedicated hockey analytics site (as far as he knows, anyway). There he pioneered the idea of using a player's plus/minus components to estimate player ice time and developed widgets such as the Disciplined Aggression Proxy to make use of the NHL's then-new RTSS stats. He has also written for the Hockey Zone Plus, Fantasy Hockey and *Hockey Prospectus* sites, and at the last-named site he developed the Projectinator, a tool designed to predict the future value of draft-eligible players based on their junior and college statistics.

Iain considers himself at least as much a historian as an analyst of hockey (he was the first person to research and publish Hall of Famer Dan Bain's career statistics), and he now focuses on applying analytic methods to historical data, seeking to excavate meaning from the misshapen heaps of archaic information. A member of the Society for International Hockey Research (SIHR) since 2000, he has served on the editorial committee of the *Hockey Research Journal* since 2006 (including four years as editor) and has had eight articles published in its pages. One such article was reprinted in *Pucklore, the Hockey Research Anthology.* Iain also contributed to two editions of the *Hockey Prospectus* annual, for 2010–11 and 2011–12.

"The convergence of hockey history and analysis" is the tagline of Iain's blog, *Hockey Historysis,* where he now focuses his research efforts. In early 2014, he published *On His Own Side of the Puck,* a book discussing the origins and development of early ice hockey rules, which was previously a subject characterized by a great deal of unknowns and uncorroborated assertions. Coming soon, Iain will publish *A Nor'west Blizzard,* a book on the history of hockey in Manitoba from 1890 to 1925.

TOM AWAD

Email: tom.awad@gmail.com

There is likely no single person on Earth who likes numbers more than Tom Awad. He started tabulating hockey statistics by hand at age 11 while watching the Montreal Canadiens games in his parents' basement. Even back then, he found the numbers to be as fascinating as the games. His eureka moment came when he came across a friend's copy of *Total Baseball* and discovered Pete Palmer's linear weights. He immediately designed something similar for hockey and tested it by transcribing the full statistics of the 1995–96 season by hand. Goals versus threshold (GVT) was born.

Tom may be best known for GVT, but over the years he has performed statistical analysis of the draft, shot quality, player usage, goaltending, and much more. He developed the projection system known as VUKOTA, which was found by David Staples of the *Edmonton Journal* to be the most accurate of seven systems for two consecutive seasons. He is a founding member of *Hockey Prospectus* and his analyses have also been published in ESPN Insider, *Arctic Ice Hockey,* and Montreal's *Journal Metro.*

When not analyzing hockey numbers, Tom can often be found analyzing other numbers in his job as an electrical engineer in Montreal or discussing mathematics, physics, or the relative merits of Sauron and Unicron with his children, David (aged nine) and

Karina (aged seven). His wife, Marisa (age redacted), sadly does not share his love of numbers, but nobody's perfect.

JOSH SMITH, ILLUSTRATOR
Twitter: @joshsmith29

Josh is a freelance writer, illustrator, and photographer, a former radio host, and a one-time (literally, just once) stand-up comedian from Long Beach Island, New Jersey. His work has been featured on Yahoo! Sports, the Hockey Writers, the Whistle, and other sites. He also runs Scouting the Refs, a site dedicated to hockey officiating.

BACK COVER MATERIAL

"Rob Vollman is one of the pioneers in the hockey analytic community. His vision and perspective on hockey has created many convincing discussions in the evaluation of today's teams and players. Rob's work is highly respected throughout the hockey world."

–JIM NILL, GM, DALLAS STARS

"*Stat Shot* does what many say is an impossible task; it makes the world of hockey analytics not just accessible, but fun. It's like math delivered with a wink and a smile."

–DAMIEN COX, JOU RNALIST/BROADCASTER

"Rob Vollman is one of the leading voices in hockey analytics, and I've learned a lot from him."

–JAMIE MCLENNAN, TSN

"Vollman's work is both groundbreaking and practical-he makes sense of hockey analytics for everyone."

–KELLY HRUDEY, *HOCKEY NIGHT IN CANADA* AND ROGERS SPORTSNET

"Analytics are here to stay in hockey and so is Rob Vollman, who gives us all something to think about with his original thinking in *Stat Shot*."

–BOB MCKENZIE, TSN HOCKEY INSIDER

"With *Stat Shot,* Vollman has found a way to take readers into deep water in hockey analytics in an easy and at times humorous way. Considering where analysis of the game is heading, this is a must-read for those who want to join the conversation and dig deeper into what's really happening inside the game."

–JEFF MAREK, SPORTSNET

"Nobody does a better job of breaking down complicated analytics for a mass audience than Rob Vollman. This isn't just a must-read for hockey fans, it should have a place on the shelves of every NHL front office."

–CRAIG CUSTANCE, ESPN.COM **AND NHL INSIDER**

***

ADVANCED STATS GIVE HOCKEY'S POWERBROKERS AN EDGE, and now fans can get in on the action. *Stat Shot* is a fun and informative guide to what analytics say about team building, a player's junior numbers, faceoff success, save percentage, the most one-sided trades in history, and everything you ever wanted to know about shot-based metrics. An invaluable supplement to traditional analysis, *Stat Shot* can be used to test the validity of conventional wisdom and to gain insight into what teams are doing behind the scenes-or maybe what they should be doing.

Made in the USA
Monee, IL
14 February 2023

27735275R00275